Advancing Healthcare Through Personalized Medicine

Advancing Healthcare Through Personalized Medicine

Priya Hays, PhD
Cepheid Corporation
Sunnyvale, CA

CRC Press
Taylor & Francis Group
Boca Raton London New York

CRC Press is an imprint of the
Taylor & Francis Group, an **informa** business

CRC Press
Taylor & Francis Group
6000 Broken Sound Parkway NW, Suite 300
Boca Raton, FL 33487-2742

© 2017 by Taylor & Francis Group, LLC
CRC Press is an imprint of Taylor & Francis Group, an Informa business

No claim to original U.S. Government works

Printed in Canada on acid-free paper

International Standard Book Number-13: 978-1-4987-6708-8 (Paperback)

Visit the Taylor & Francis Web site at
http://www.taylorandfrancis.com

and the CRC Press Web site at
http://www.crcpress.com

For Shreyas, Tejas, Sahil, and Rohan
Four Treasures

Contents

Preface

Medicine as a field in the early twentieth century was based on the humoral hypothesis, positing that an excess or deficiency of any of four distinct bodily fluids in a person—known as humors or humours—directly influences his or her temperament and health. The humoral hypothesis was based on early Greek and Roman physiology. The humoral hypothesis was replaced in the 1900s by the sciences of microbiology, biochemistry, and biology. A paradigm shift took place that entirely transformed medicine into the way it is practiced today.

Similarly, a new paradigm shift at hand in healthcare is based on the emerging field of personalized medicine, a disruptive innovation that will profoundly change our thinking of medicine. Personalized medicine concerns targeted patient treatment; in other words, rather than pursuing a "one-drug-fits-all" approach, personalized medicine providers in healthcare focus on tailoring drugs to the specific needs of each patient through stratification. Personalized medicine focuses on using biomarkers identified through genetics, proteomics, and metabolomics for prevention and therapy. With the advent of genomic technologies, personalized medicine will have the capability to proactively predict and diagnose disease efficiently. Personalized medicine has already led to significant advances in cures for cancer, including treatment for breast, colon, lung, and blood cancers. It has the potential to revolutionize healthcare, as have other "revolutions" in medicine that have improved the quality of life for Americans, for example, insulin for diabetics and the polio vaccine for all individuals. It is crucial that people become familiar with personalized medicine, the science behind it, its economic effects, its effects on patients, and its overall implications for society.

As medicine is practiced today based on the pathophysiology of disease, clinicians focus on acute episodes of chronic disease. Chronic illness grows over time: the genome and environment lead to development of disease, once reaching a point where it will manifest clinically. The more the patient waits, the less reversibility and greater costs there are, constituting very little awareness into the root causes of chronic disease and "backwards-looking medicine." Many physicians were looking at improving this paradigm and creating fundamentally new ways of practicing medicine. Personalized medicine arose under this backdrop of developments. Emerging technologies would now target disease based on early molecular detection leading to earlier clinical detection. Medicine began to evolve in this direction as early as 2002, from reactive medicine to patient-driven proactive medicine, with coinvolvement of patients and physicians in healthcare.

I became interested in personalized medicine by attending my first conference on the subject, the 2013 Personalized Medicine World Conference held by Silicom Ventures. Funding from the company I was working at at the time, Affymetrix, allowed me to attend the conference. I was fascinated by the advances, treatments, and changes to medicine brought about by the implementation of personalized medicine. I learned about companion diagnostics, the economics of personalized medicine, and recent Food and Drug Administration advances to change clinical trials brought about by personalized medicine (used interchangeably with precision medicine).

This interest did not come naturally as the result of my work. I have written two books: *Molecular Biology in Narrative Form*, based on the dissertation I conducted in the Department of Literature at University of California, San Diego, and *Science, Cultural Values and Ethics*, based on my postdoctoral research fellowship in hematological research at Dartmouth Medical School. However, these books were more in the areas of biomedicine and society, rather than reviews of the state of scientific advances. Through the course of my academic career, I published articles in such prestigious journals as *L'Esprit Createur*, a preeminent French studies journal, and the *Bulletin of Science, Technology and Society*, a tour de force in presenting articles on science, technology, and society studies. I even had an article on the state of interdisciplinarity in the journal *Interdisciplinary Literary Studies*, extolling the intellectual accomplishments of the humanities.

Yet, my work was considered abstract and theoretical. After observing members of my family suffering from Huntington's disease, a fatal neurodegenerative illness, I turned a page.

After hearing about clustered regularly interspaced short palindromic repeats (CRISPR) technologies and how gene editing had the potential to cure or reverse inherited disease, I began to see the merits of biomedical progress and how it can affect patient lives, and wanted to devote my writing efforts to making a difference in patient lives and helping physicians navigate through the complex maze of personalized medicine. Thus, this book is the outcome of numerous interviews, conference presentations, patient narratives, research, and painstaking literature surveys. Even though the experience of writing this book has challenged my ideals and attitudes, I still remain positive about the outcomes that precision medicine can bring about for patients, physicians, and the public in general.

The gains in healthcare through personalized medicine have been varied, and the current state of affairs remains as follows. Technology has been breathtaking through the development of proteomics and metabolomics and the capacity to analyze big data moving rapidly. We have made some advances in predictive and diagnostic tools in terms of companion diagnostics. Cancer therapeutics have constituted a paradigm shift; however, other areas of diseases have not progressed far enough. Reimbursement remains a barrier, not supporting the culture of proactive medicine. Chronic disease mitigation also receives poor grades. Each of these areas is addressed in this book, accompanied by a fair assessment of progress and challenges.

By chronicling the most recent advances in personalized medicine, *Advancing Healthcare through Personalized Medicine* aims to inform healthcare providers, scientists, industry and government leaders, and members of the business community, and perhaps even the general public, of the coming revolution in healthcare. Beginning with President Barack Obama's Precision Medicine Initiative (now also known as the All of Us Research Program), which he unveiled at the 2015 State of the Union, to look at targeted treatment approaches such as Gleevec, a widely used cancer drug, the book delves deeply into the promise of personalized medicine, how it will drive innovation in biomedicine, and the challenges to its implementation. Personalized medicine will have broadly reaching effects on society, both financially and ethically, and the economic and moral implications of personalized medicine remain at the forefront of the stakeholder interviews, discussions, examples, case studies, and patient narratives in the book. The book uses jargon-free descriptions and describes how targeted genomic medicine and its accompanying companion diagnostics will constitute the future of healthcare. In short, I explain how the current healthcare crisis can be mitigated through the emergence of personalized medicine.

As I have found through my research, personalized medicine means many different things to people depending on position—clinician, patient, payer, or drug developer, among others—and within these roles, there are even more perspectives. Public ventures such as the Precision Medicine Initiative, which aims to have a cohort of volunteers sequenced in a shared database, still compete with private ventures such as Craig Venter's Human Longevity Institute, which aims to determine any individual's omics, resulting in a secured database. Privacy issues remain paramount, along with data integrity and storage. Personalized oncology has made the greatest gains, but cardiovascular disease and the central nervous system are quietly progressing. The value of companion diagnostics still remains to be determined, and venture capital is still not incentivized completely toward it. Gene panels remain active for tumor biology, but there are advocates for next-generation sequencing. These issues are discussed in more detail in this book. Perhaps I have raised more questions from this monograph than provided answers, but I hope to have enlightened the reader on one of the greatest advances in the twenty-first century: personalized and precision medicine.

Priya Hays
San Jose, California

Acknowledgments

This book on personalized medicine has benefitted greatly from the contributions of many people through interviews, presentations, case studies, patient narratives, and scientific and medical articles. Through conferences on personalized medicine, I have collected information from presentations to present the latest in leading-edge treatments for prevention and cures of diseases through personalized medicine. These presenters served as outstanding sources for state-of-the-art material on genomic technologies, pharmacogenomics, and basic research in personalized medicine. Interviews took place with individuals who are at the forefront of implementing personalized medicine. Many others facilitated these interviews by supporting my attendance at conferences and by suggesting people to interview.

I wish to give special thanks to the following individuals, whose expertise make up this book and who have been instrumental in the advancement and implementation of personalized and precision medicine: Edward Abrahams and his staff at the Personalized Medicine Coalition, whose exemplary efforts in the implementation of personalized medicine remain at the forefront of biomedicine; Angel Anaya; Retta Beery, who provided firsthand insight into the medical journey of her extraordinary twin son and daughter; Carol Berry of Asuragen for her information on molecular diagnostics of thyroid cancer; Trevor Bivona of the University of California, San Francisco, who provided an informative interview on the gains of personalized medicine; Erwin Bottinger of Mount Sinai School of Medicine, who provided his time on the efforts of Mount Sinai in biobanking; Wylie Burke of the University of Washington, a premier bioethicist; Robert Califf of the Food and Drug Administration for his presentation on the regulatory landscape; Michael Christman of the Coriell Institute; Joshua Cohen of the Tufts Center of Drug Development for his outstanding review and helpful comments, from which the book benefited greatly; Raymond DuBois; Bob Dvorak, whose outstanding insights into personalized medicine drugs inform this book; Gianrico Farrugia of the Mayo Clinic, whose terrific and thorough comments insisting on clinical data and evidence noticeably improved this book; Matthew Ferber of the Mayo Clinic for his education on next-generation sequencing; Mark Gerstein of Yale Computational Biology for his educational lecture on genetic privacy; Gary Ginsburg of Duke for his tremendous efforts in articulating the gains of personalized medicine through his lectures and work; Hank Greely of Stanford University, whose influence on the bioethics of personalized medicine is monumental; Eric Green of the National Human Genome Research Institute, who provided an introduction to genomics at the Mayo Clinic; Houda Hachad of Translational Software, whose interview provided much insight into pharmaogenetics; Clifford Hudis of Sloan Kettering Cancer Center; Kathy Hudson of the National Institutes of Health for her informative lecture on the Precision Medicine Initiative; Paul Hudson of AstraZeneca for an outstanding conclusion to the Partners Conference; Howard Jacob of the Medical College of Wisconsin, whose talk at the Mayo Clinic was insightful; Kiley Johnson, formerly of the Mayo Clinic, for her information on genetic counseling; Eric Klee of Mayo Clinic; Bartha Knoppers of the Global Alliance for Genomics and Health; Raju Kucherlapati of Harvard Medical School, who serves as an exemplary leader in precision medicine; David Ledbetter of Geisinger Health Systems; Lawrence Lesko; Elaine Lyon of the University of Utah; Noralane Lindor of the Mayo Clinic for her excellent overview of diagnostic odysseys; Jen Madsen of Arnold & Porter LLP; Elaine Mardis of Washington University, whose groundbreaking work in the area of personalized medicine informs this book; Barbara McAneny for her insight into precision medicine in the community; Robert McMorris of the Mayo Clinic; Mark Monane of Corus Dx, a company that will revolutionize cardiovascular care; Don Morris of Archimedes; Bob Nellis of the Mayo Clinic for his generosity and funding; Parisa Mousavi for her outstanding course on molecular diagnostics; Gitte Pederson of Genomic Expression for her talk on her innovative company; Tea Pemovska of FIMM for her outstanding presentation on systems biology and oncology; Gloria Peterson of the Mayo Clinic on clarifying the role of biobanks; Kim Popovits, CEO of Genomic Health, a leader in molecular diagnostics; Matt Posard, formerly of Illumina, a game changer whose work provides the backdrop for the book's conclusion; Stephen Quake of Stanford for his lecture at the Mayo Clinic and pioneering work on noninvasive sequencing; Ken Ramos of the University of Arizona; Robert Green and Heidi Rehm of Partners Healthcare at Harvard Medical School, who provided an informative interview on how their group is making genomic information available to the medical community through their MedSeq project; Mary Relling; David Resnick for his insight into biotech patents; Chris Riedel, formerly of Hunter Labs in Campbell, California, for his interview, from whose passion and insight personalized medicine will benefit enormously; Dan Roden of Vanderbilt University, a brilliant physician who generously lent his time to informing me about pharmacogenetics; Laura Lyman Rodriguez of the National Human Genome Research Institute for promulgating the Precision Medicine Initiative's efforts in protecting privacy; Don Rule, an exceptional leader in making pharmacogenetics available to clinicians through his company Translational Software, for providing an interview on his efforts; Michael Snyder of Stanford University, an inspired scientist and great man, for providing the first interview for the book and my introduction to personalized medicine; Ralph Snyderman of Duke University, a superb leader in the field of personalized medicine, for his exceptional lectures, which I attended at various personalized medicine conferences; Stephen Stahl of the University of California, San Diego, for his clear insight into psychiatry and the role of personalized medicine; Dietrich Stephan of Silicon Valley Biosystems; Mark Trusheim of the Massachusetts Institute of Technology for his extraordinary insights into the economics of personalized medicine and the value to healthcare that personalized medicine will bring; Richard Watts of Qiagen; Richard Weinshilboum of the Mayo Clinic for his work in spearheading pharmacogenomics into clinical care, which furthered this manuscript significantly; Frank Witney, the CEO of Affymetrix, for making me aware of the National Academies Initiative in precision medicine; and Dara Wright of eBiosciences, a great manager who gave me my introduction to the biotechnology world and provided funding to the first personalized conference I attended.

I must also recognize the Mayo Clinic, Partners Healthcare of Harvard Medical School, Arrowhead Publishers, and Silicom Ventures, as well as the preeminent authors of the informative biomedical articles I encountered during my research. A special thanks for Naomi Wilkinson and her outstanding efforts in getting the book published and my publisher CRC Press/Taylor & Francis Group.

My best regards and appreciation for my coworkers at Cepheid, and a shout-out to my one-in-a-million husband, Ezra, the best thing that ever happened to me.

This book is dedicated to my nephews Shreyas, Tejas, Sahil, and Rohan, whose love and affection have brought great joy to my world.

Finally, this book is written for the patients, families, and communities who have faced life's impediments because of disease, illness, and sickness, and for physicians who have met with frustration in traditional trial-and-error medicine. If this book can provide hope to them in any form through personalized and precision medicine, then its main purposes have been met.

CHAPTER 1

Introduction: Biomedical innovation and policy in the twenty-first century

The question is not whether personalized medicine is here to stay; it is how fast is it going to be implemented.

Raju Kucherlapati, PhD
Professor of Medicine and Paul C. Cabot Professor of Genetics
Harvard Medical School

The development of Xalkori, also known as crizotinib, a small molecular inhibitor encoded by the ALK gene targeted to treat non-small-cell lung cancer (NSCLC), began several years ago. Analysis of a cDNA library Japanese patient with lung adenocarcinoma identified a novel fusion between the EML4 and ALK genes with the ability to transform 3T3 fibroblasts. Analysis of a series of biopsies from NSCLC patients revealed that ~5% of patients carry this fusion protein (Ranade et al., 2014). After this initial finding in 2007, crizotinib was discovered to be an effective targeted therapy for patients whose NSCLC tumors harbored the ALK gene. According to Ranade et al., it caused tumors to shrink or stabilize in 90% of 82 patients carrying the ALK fusion gene, and tumors shrank at least 30% in 57% of people treated. These promising clinical results led to phase II and a phase III trials, which selectively enrolled NSCLC patients with ALK fusion genes. Astonishingly, within four years of the initial publication by Soda et al. (2007), the Food and Drug Administration (FDA) approved crizotinib for the treatment of certain late-stage (locally advanced or metastatic) NSCLC patients whose tumors had

ALK fusion genes, as identified by a companion diagnostic that was approved simultaneously with the drug, as Ranade et al., noted.

When the FDA approved the cancer drug known as Gleevec in 2001 for treatment against chronic myeloid leukemia (CML), the agency did so after one of the shortest drug review processes on record. Novartis, the maker of Gleevec, also known as imatinib mesylate, had sponsored the clinical trials behind the drug. The FDA approved the drug within 10 weeks of reviewing three separate studies on 1000 patients. On May 10, 2001, at the press conference announcing the FDA approval of Gleevec, Health and Human Services (HHS) secretary Tommy Thompson declared:

Today I have the privilege of announcing a medical breakthrough. Like most scientific breakthroughs, this one is *not* sudden, nor does it stand alone. Rather, like most scientific advancement, it is a culmination of years of work and years of investment, by many people in many different institutions, and even in different fields of medicine. We are here to announce one dramatic product of all those efforts. But we believe many more products will follow, based on years of scientific groundwork. So this is the right time to acknowledge those efforts, to recognize that our investments in research are paying off, and to praise the teamwork that has brought us here. It's also the right time to talk about what this can mean for our future—a future that promises a new level of precision and power in many of our pharmaceutical products. Today the Food and Drug Administration has approved a new drug, Gleevec, for treatment of chronic myeloid leukemia—or CML. Let me just say that it appears to change the odds dramatically for patients. And it does so with relatively low occurrence of serious side effects.

With the details of the effectiveness of Gleevec and its implications, Thompson's announcement quickly gained the news media's attention. CNN cycled the story every half hour throughout the day of the press conference. The Associated Press wrote and updated the story several times, and the news made the front page of newspapers nationwide. In the weeks following the announcement, extraordinary coverage was given to Gleevec and its effects on cancer, including a cover story in *TIME* magazine (May 28, 2001) and reports in the *New York Times*, *USA Today*, and *Newsweek*.

Gleevec was proven to be effective not just against CML, but also against another cancer, gastrointestinal stromal tumor (GIST). Three days after Thompson's press conference, during the annual meeting of the American Society of Clinical Oncology in San Francisco, Dr Charles Blanke announced that the so-called "leukemia pill" had stunning results against GIST (https://liferaftgroup.org/wp-content/uploads/2012/09/May2001newsletter.pdf).

According to a website for a GIST patient advocacy group a dozen years after its approval, Gleevec is a relatively unknown pill. Why all the attention focused on one orange pill against two relatively rare cancers (CML affects 4500 patients annually, while GIST is even rarer)? Although primarily addressing CML rather than GIST, Thompson broadly answered the question at the HHS press conference: Gleevec is targeted therapy—it kills leukemia cells while sparing normal white blood cells. Unlike other more strenuous chemotherapy regimens, Gleevec has relatively few side effects. Gleevec targets the signal in the cell that causes cancer, acting as a molecular switch. Gleevec is now the prototype of cancer drugs, and cancer research laboratories around the world are trying to mimic the effects of Gleevec on other types of cancers.

As reported by the National Cancer Institute, most of the 4500 Americans diagnosed with CML each year are middle-aged or older, although some are children. In the first stages of CML, most people do not have any symptoms of cancer, as disease progresses slowly. Bone marrow transplantation in the initial chronic phase of the disease is the only known cure for CML. However, many patients are not young or healthy enough to tolerate transplantation; of those expected to tolerate transplantation, many cannot find a suitable donor, and the procedure can cause serious side effects or death. For these patients, treatment with the drug interferon alfa, introduced about 20 years prior to Gleevec, may produce remission, restoring a normal blood count in up to 70% of patients with chronic-phase CML. If interferon alfa is ineffective or patients stop responding to the drug, the prognosis is generally bleak.

Gleevec has produced higher remission rates in three short-duration, early-phase clinical trials. In the results of one clinical trial, reported in April 2001 in the *New England Journal of Medicine*, Gleevec restored normal blood counts in 53 out of 54 interferon-resistant CML patients, a response rate rarely seen in cancer with a single agent. Fifty-one of these patients were still doing well after a year on the medicine, and most reported few minor side effects. Imatinib mesylate was invented in the late 1990s by scientists at Ciba-Geigy (which merged with Sandoz in 1996 to become Novartis), in a team led by biochemist Nicholas Lydon, and its use to treat CML was driven by oncologist Brian Druker of Oregon Health & Science University. Druker led the clinical trials confirming its efficacy in CML (Gambacorti-Passerini, 2008). The scientific story of Gleevec, which became known as targeted therapy, a medical breakthrough that was a result of years of research, heralds back to the discovery of the BCR-ABL mutation in chromosomes 9 and 22 by Janet Rowland at the University of Chicago, and the pioneering work of researchers at Johns Hopkins University who discovered that cancer cells harbor this mutation. Gleevec is targeted therapy, designed to attack cells with this BCR-ABL mutation (also known as translocation, when pieces of chromosomes detach from one or more chromosomes and move to another chromosome). This BCR-ABL mutation affects a growth pathway in the cell known as the tyrosine kinase pathway, which leads to a cancerous state.

For the first time, cancer researchers now have the necessary tools to probe the molecular anatomy of tumor cells in search of cancer-causing proteins," said Richard Klausner of the National Cancer Institute. "Gleevec offers proof that molecular targeting works in treating cancer, provided that the target is correctly chosen. The challenge now is to find these targets (http://www.cccbiotechnology.com/WN/SU/gleevecnews.php).

There are hundreds of known mutations for cystic fibrosis (CF), an inherited disease that affects the lung leading to complications such as pneumothorax. In 2012, the FDA approved a new therapy for CF called Kalydeco (known generically as ivacaftor), which was approved for patients with a specific genetic mutation—referred to as the G551D mutation—in a gene that is important for regulating the transport of salt and water in the body. The G551D mutation is responsible for only 4% of cases in the United States (approximately 1200 people). In these patients, Kalydeco works by helping restore the function of the protein that is made by the mutated gene. It allows a proper flow of salt and water on the surface of the lungs and helps prevent the buildup of sticky mucus that occurs in patients with CF and can lead to life-threatening lung infections and digestive problems.

The FDA's profile of personalized medicine chronicles the development of Herceptin (Simoncelli, 2013):

The story of trastuzumab (Herceptin, made by Genentech, Inc.) began with the identification by Robert Weinberg in 1979 of "HER-2," a gene involved in multiple cancer pathways. Over the next two decades, UCLA researcher Dennis Slamon worked to understand the link

between HER2 and specific types of cancer. Slamon observed that changes in the HER2 gene caused breast cancer cells to produce the normal HER2 protein, but in abnormally high amounts.

Overexpression of the HER2 protein appeared to occur in approximately 20%–25% of breast cancer cases, and seemed to result in an especially aggressive form of the disease. These observations made it clear that HER2 protein overexpression could potentially serve as both a marker of aggressive disease as well as a target for treatment. In May 1998, before an audience of 18,000 attendees of the annual meeting of the American Society for Clinical Oncology (ASCO), Slamon presented evidence that Herceptin, a novel antibody therapy he had developed in collaboration with researchers at Genentech, was highly effective in treating patients with this extraordinarily aggressive and intractable form of breast cancer. What was so revolutionary about Herceptin was that it was the first molecularly targeted cancer therapy designed to "shut off" the HER2 gene, making the cancerous cells grow more slowly and without damaging normal tissue. This precision also meant that patients taking the new treatment generally suffered fewer severe side effects as compared with other cancer treatments available at that time.

In September 1998, FDA approved Herceptin for the treatment of HER2 positive metastatic breast cancers. On that same day, the Agency granted approval to DAKO Corp for HercepTest, an in vitro assay to detect HER2 protein overexpression in breast cancer cells.

Four stories, four drugs. Each of these highlights certain aspects of personalized medicine and positive lessons learned: the discovery of driver mutations that drugs could target, rapidly facilitated clinical trials that lessen FDA approval time for breakthrough drugs, and codevelopment of drug and companion diagnostics that lead to effective predictive treatment for patients. Raju Kucherlapati's statement is telling of the coming revolution in biomedicine ahead of us.

One of the most remarkable changes has occurred in the landscape of clinical trials in the wake of personalized (physician approaches) and precision (pharma approaches) medicine. By identifying driver mutations in heterogenous tumors that could serve as targets for therapy, drug companies could save time and money in drug development by "designing small but highly effective trials targeted to those patients more likely to benefit from the therapy" (Ranade et al., 2014). In a study led by researchers at Weill Cornell Medical College in 2015, results identified 684, or 8%, of eligible trials as precision cancer medical trials that were significantly more likely to be phase II and multicenter; involved breast, colorectal, and skin cancers; and required 38 unique genome alterations for enrollment. The proportion of precision cancer medicine trials compared with the total number of trials increased from 3% in 2006 to 16% in 2013 (Roper et al., 2015).

> In July 2015, oncologists will start enrolling patients in a clinical trial with 20 or more arms, each testing different agents against different molecular targets and each including patients with different cancers. In design, the trial itself couldn't be more different from the classic clinical trial.
>
> Instead of focusing on one cancer, as trials have for decades, the National Cancer Institute's NCI-MATCH (Molecular Analysis for Therapy Choice) trial will include patients with any solid tumor or lymphoma who have one of many genomic abnormalities known to drive cancer. Patients will be matched with a targeted agent that has shown promise against their abnormality, regardless of what cancer they have. Known as a basket trial, the new design highlights the rapidly growing number of potential targets and agents in oncology and the urgency of finding more efficient ways to evaluate them in trials (McNeil, 2015).

According to a review published by the National Cancer Institute, recent advancements in cancer biomarkers and biomedical technology have begun to transform the fundamentals of cancer therapeutics and clinical trials through innovative adaptive trial designs. The goal of these studies is to learn not only if a drug is safe and effective, but also how it is best delivered and who will derive the most benefit (Heckman-Stoddard and Smith, 2014). Heckman-Stoddard and Smith (2014) cite two trials under way: I-SPY and BATTLE:

> I-SPY 1 is a neoadjuvant trial of women with locally advanced breast cancer, which are assessed for estrogen receptor (ER), progesterone receptor, human epidermal growth factor 2 (HER2), and Mammaprint, a 70-gene predictive signature of distant recurrence prior to treatment (or randomization). The trial evaluates molecular biomarkers of treatment and response and breast imaging to guide "adaptive" (i.e., subsequent optimal treatments). Initial studies were used to develop and validate optimal metrics of treatment response in I-SPY1.
>
> In I-SPY 1, chemotherapy was administered before surgery, and biomarkers were compared with tumor response on the basis of magnetic resonance imaging (MRI), pathologic residual disease at the time of surgical excision, and 3-year disease-free survival. The study found that pathologic complete response (pCR), defined as no invasive tumor present in either the breast or axillary lymph nodes, differed by molecular subset; hormone receptor-positive/HER2-negative carcinomas were associated with the lowest pCR (9%) and hormone receptor-negative/HER2-positive had the highest pCR (45%). I-SPY 1 also indicated that pCR was predictive of recurrence free survival within a molecular subset. The study showed that MRI volume was the best predictor of residual disease after chemotherapy. This study established the infrastructure to integrate biomarkers and imaging with shared methods and real-time access to study data which will be leveraged for I-SPY 2....
>
> The phase II Biomarker-integrated Approaches of Targeted Therapy for Lung Cancer Elimination (BATTLE)

program is a second example of a clinical trial to determine regimes for precision or personalized medicine. Biomarkers have emerged as an important factor in planning treatment for non-small cell lung cancer (NSCLC) because of knowledge that specific epidermal growth factor receptor (EGFR) mutations lead to improved outcomes with EGFR tyrosine kinase inhibitors (TKI). The BATTLE program consists of an umbrella trial plus four phase II protocols. These phase II protocols used agents directed against promising molecular targets at the time the study began in 2005. The targets included EGFR (erlotinib), KRAS/BRAF (sorafinib), retinoid-EGFR signaling (bexarotene and erlotinib), and vascular endothelial growth factor receptor (VEGFR) (vantetanib). The primary endpoint of the study was the 8-week disease control rate (DCR) defined as complete or partial response or stable disease via Response Evaluation Criteria in Solid Tumors (RECIST). A 30% DCR in similar patients was used as a control, with treatment efficacy defined as a greater than 80% probability of achieving greater than 30% DCR.

Patients enrolled in the umbrella trial underwent tumor biopsy and biomarker analysis for 11 biomarkers: mutations in EGFR, KRAS, and BRAF; copy numbers of EGFR and the Cyclin D1 gene (CCND1); and protein expression level of VEGF, VEGF-2, RXRs α, β, and γ, and Cyclin D1. The biomarker analysis was completed with day 14 biopsy; patients and investigators were blinded to biomarker results until the patient went off study. The results of the biomarker analysis were used to classify patients into one of five groups: (1) EGFR mutation and/or amplification; (2) KRAS or BRAF mutation; (3) VEGF and/or VEGF-2 overexpression; (4) RXRs α, β, and γ, and/or Cyclin D1 overexpression and/or CCND1 amplification; or (5) negative for biomarker panel. Those patients who were positive for more than one marker were

assigned to a treatment group based on the marker with highest predictive value. During the first part of the study patients were enrolled randomly to each of the four phase II studies except for patients with prior erlotinib treatment who were excluded from the erlotinib-containing study arms. The results of this randomized portion of the study were used to assess the association between a given marker group and disease control. For example, patients with an EGFR mutation and/or amplification had a certain probability of disease control with each of the treatment regimens. For the second part of the trial this probability was incorporated into a Bayesian adaptive algorithm to randomly assign patients to an optimally predicted treatment arm. The probability of disease control was updated throughout the trial based on accumulating data.

Efficacy outcomes for these biomarker-driven trials have also been demonstrated in one study published in the *Journal of the National Cancer Institute* led by investigators at the Moores Cancer Center at the University of California, San Diego:

In order to ascertain the impact of a biomarker-based (personalized) strategy, outcomes were compared between US Food and Drug Administration (FDA)-approved cancer treatments that were studied with and without such a selection rationale. The results: fifty-eight drugs were included (leading to 57 randomized [32% personalized] and 55 nonrandomized trials [47% personalized]). Trials adopting a personalized strategy more often included targeted oral and single agents and more frequently permitted crossover to experimental treatment. In randomized registration trials (using a random-effects meta-analysis), personalized therapy arms were associated with higher relative response rate ratios compared with nonpersonalized trials. Analysis of experimental arms in all 112 registration trials (randomized and nonrandomized) demonstrated that personalized therapy was associated with higher response rate.

The study authors conclude that "a biomarker-based approach was safe and associated with improved efficacy outcomes in FDA-approved anticancer agents" (Fontes Jardim et al., 2015).

These precision treatments and the outcomes of these clinical trials, or the development of them over time, have led to awareness among sectors of the community that personalized medicine would have a definitive impact on biomedicine. On January 21, 2015, President Barack Obama unveiled the Precision Medicine Initiative in his State of the Union to members of Congress. "I'm launching a new nationwide Precision Medicine Initiative to bring us closer to curing diseases like cancer and diabetes," he stated. According to the mission statement of the Precision Medicine Initiative, precision medicine

> Enable[s] a new era of medicine through research, technology, and policies that empower patients, researchers, and providers to work together toward development of individualized treatments. The future of precision medicine will enable health care providers to tailor treatment and prevention strategies to people's unique characteristics, including their genome sequence, microbiome composition, health history, lifestyle, and diet. To get there, we need to incorporate many different types of data, from metabolomics (the chemicals in the body at a certain point in time), the microbiome (the collection of microorganisms in or on the body), and data about the patient collected by health care providers and the patients themselves. Success will require that health data is portable, that it can be easily shared between providers, researchers, and most importantly, patients and research participants.

The "omics" era has entered.

The fact sheet on the Precision Medicine Initiative reads like a primer on personalized medicine:

Launched with a $215 million investment in the President's 2016 Budget, the Precision Medicine Initiative will pioneer a new model of patient-powered research that promises to accelerate biomedical discoveries and provide clinicians with new tools, knowledge, and therapies to select which treatments will work best for which patients. Most medical treatments have been designed for the "average patient." As a result of this "one-size-fits-all-approach," treatments can be very successful for some patients but not for others. This is changing with the emergence of precision medicine, an innovative approach to disease prevention and treatment that takes into account individual differences in people's genes, environments, and lifestyles. Precision medicine gives clinicians tools to better understand the complex mechanisms underlying a patient's health, disease, or condition, and to better predict which treatments will be most effective. Advances in precision medicine have already led to powerful new discoveries and several new treatments that are tailored to specific characteristics of individuals, such as a person's genetic makeup, or the genetic profile of an individual's tumor. This is leading to a transformation in the way we can treat diseases such as cancer. Patients with breast, lung, and colorectal cancers, as well as melanomas and leukemias, for instance, routinely undergo molecular testing as part of patient care, enabling physicians to select treatments that improve chances of survival and reduce exposure to adverse effects (https://www.whitehouse.gov/the-press-office/2015/01/30/fact-sheet-president-obama-s-precision-medicine-initiative).

Investments to launch this initiative include funding to the National Institutes of Health (NIH), FDA, and National Cancer Institute to identify genomic drivers in cancer. Key objectives include more and better treatments for cancer, the creation of a voluntary cohort of patient data (participants will be involved in the design of the initiative and will have the opportunity to contribute diverse sources of data, including medical records; profiles of the patient's genes, metabolites [chemical makeup], and micro-organisms in and on the body; and environmental and lifestyle data), and the protection of privacy.

Kathy Hudson of the NIH declares that the Precision Medicine Initiative is one small part of precision medicine. The PMI will consist of building a cohort of 1 million volunteers to facilitate building effective preventive strategies, diagnostics, and treatments. Hudson and stakeholders in the PMI would like this program to treat participants as partners rather than subjects, and look beyond the genome to integrate other sources of knowledge about health, including diet and environment, to develop quantitative estimates of risk for a range of diseases by integrating environmental exposures and genetic factors. Concerning this cohort study, in a survey of public opinion, 79% agree that a cohort should be done and 54% would participate, with 83% agreeing to receive the results of the survey. One of the concerns of this cohort study is the return of data and who should have access to it. According to Hudson, participants agreed that data can be accessed broadly for research purposes. Francis Collins, director of the NIH, has set up a working group for implementing this cohort study with a steering committee, according to Hudson.

The advent of precision medicine heralds back to the days when the first targeted treatment approaches were just appearing on the horizon. Xalkori, Gleevec, Kalydeco, and Herceptin are part of a new wave of targeted therapies. Gleevec serves as a significant example of personalized medicine because it was one of the first therapies designed to attack cancer by confronting the molecular causes of cancer.

These targeted treatment approaches herald a new wave of healthcare: personalized medicine. Personalized medicine means many things to many people, such as patients, clinicians, and other healthcare workers. As they demonstrate, personalized medicine concerns utilizing the genetic variation in individuals to tailor therapies according to the constitution of each individual's DNA, for which one can identify the molecular causes of disease and genetically diagnose patients by recognizing biomarkers that reveal predisposition to

disease. Personalized medicine is characterized as both innovation and atavistic, as both elegant and convoluted, and as both revolutionary and static. Controversial nonetheless, personalized medicine is both encouraged and supported by some, and reconsidered and questioned by others. Consider a recent *Journal of the American Medical Association* article arguing for reason to ask cogent questions about the implementation of personalized medicine (Joyner and Paneth, 2015). A noted bioethicist, Donna Dickenson, has written a monograph arguing that investment in personalized medicine may come about at the expense of public health (Dickenson, 2013). The argument that personalized medicine, or personalized medicine's potential, can lead to safe and efficacious cures that in the long term may lead to less cost for the healthcare system, as evidenced by the example of Gleevec and other drugs, is under some debate. However, less scrutiny is given to the socioeconomic impact of personalized medicine, that is, what are the effects of personalized medicine on patient outcomes. That is, how will personalized medicine influence healthcare, through a confluence of networks among the biomedical and pharmaceutical industry, government regulators, and clinicians and researchers in academia and major medical centers?

The thesis behind this book, *Advancing Healthcare through Personalized Medicine*, is that a paradigm shift is taking place in biomedicine, driven by personalized medicine, pharmacogenetics and pharmacogenomics, and genetic testing. The clinic, the pharmaceutical and biotech industry, translational medicine, and basic research will undergo major transformations, and a new set of moral and ethical issues will arise with these socioeconomic changes. How will knowing that a genetic predisposition prevents Alzheimer's, heart disease, and cancer through medication and lifestyle changes improve healthcare? Sparking a debate in society is necessary to confront the serious changes and deep consequences that will ensue from pursuing personalized medicine.

Personalized medicine is partly about addressing intergenetic variation in the population. The current strategy in drug development and treatment is the "one-drug-fits-all" kind, which has its limits in safety and efficacy, and possibly contributes to high costs for the healthcare system. Personalized medicine has the potential to produce safe and effective medicines, and to lower costs and risks, including adverse drug reactions, which run rampant in the elderly population. Pharmacogenetics and pharmacogenomics examine how genetic composition affects both disease predisposition and response to therapy, and brings the promise of a new era of "personalized medicine": delivery of the right drug to the right patient at the right dose (Piquette-Miller and Grant, 2007).

Thus, personalized medicine and pharmacogenomics represent a new era of healthcare. Ideally, a medical provider can prescribe a unique therapy for a patient, basing the therapy on the provider's diagnosis. Patients can now receive unique therapies for their diseases or conditions, and genetic tests can reveal information about genomic makeup that allow for personalized care. Additionally, translational medicine centers (translational medicine, also known as bench-to-bedside medicine, "translates" discoveries from basic science laboratories into medicine for patients) have formed or are forming in major medical centers across the United States, with the intent of delivering clinical care though basic research.

As of this writing, there is the possibility that federal funding in the United States for scientific research had dropped dramatically during the period of 2011–2013. Thus, scientific grant acceptance rates of research proposals could plummet to single digits.

Biomedical science is currently at a crossroads. While the federal government cuts funding drastically for biomedical research and pharmaceutical companies are being criticized for pursuing profits over patients, some scientists and clinicians are heralding that personalized medicine and translational medicine are the future of research and treatment.

For some years now, federal funding for basic science research has been somewhat constricted. According to a study by Dorsey et al. (2010), federal funding for research has not kept up with inflation. Their conclusion:

> After a decade (1994–2003) of doubling, the rate of increase in biomedical research funding slowed from 2003 to 2007, and compared to the 2007 figure with an adjustment for inflation, the absolute level of funding from the National Institutes of Health and industry appears to have decreased by 2% in 2008 (Dorsey et al., 2010).*

A recent editorial in the *San Jose Mercury News* points to the concern over this state of affairs in funding. The opinion writers, Roger Kornberg and Scott Bruder, cite the discovery of flow cytometry as an example of a pivotal event that resulted in part from federal funding of biomedical research, correlating the economic prosperity of California and the United States as a whole with advances in research. Kornberg and Bruder argue that investment in NIH promotes long-term growth and prosperity in the national economy, particularly in the biotech sector, which innovates at a record pace. They conclude that "our nation and region are at a critical juncture. As we struggle to cope with an aging population, debilitating chronic conditions, and deadly infectious diseases, sustained and strong NIH support will be crucial to the health of California and the nation" (Kornberg and Bruder, 2011).

This editorial echoes the sentiments of many biomedical researchers in academia, where much NIH-funded research occurs. It may in fact be the mainstream view of the biomedical community and industry as well that support for academic research will lead to economic prosperity and patient cures. This support would probably translate to more funding and discoveries in commercial research, leading to more jobs and global health improvements. In short, this editorial is exemplary of efforts to increase federal funding of biomedical research on the basis of improved health and job opportunities as an antidote to the economic recession (note that unemployment was near historical highs when the editorial was published).

There is a certain level of dissonance present. With basic science threatened by imminent downsizing due to cuts, and Big Pharma and biotech in marked transition in terms of their business

---

* However, through the 21st Century Cures Act passed by Congress in December 2016 and signed by President Obama, the NIH is authorized to receive $4.8 billion over ten years and the FDA receives $500 million. The $6.3 billion act was sponsored by Republican Representative Fred Upton and was voted overwhelmingly by the Senate and House of Representatives. The bill also designates $1.8 billion in funding for Vice President Joseph Biden's Cancer Moonshot initiative for cancer research.

models, clinicians, medical researchers, and scientists are touting a new wave of medicine that offers no less than a revolution in patient diagnosis and therapy. In other words, upstream and downstream of therapy are bottlenecks, and *in media res* are growing and thriving branches of medicine. This disconnect could well impede the process of innovation in biomedicine.

One of the arguments of this book is that increased investment in personalized medicine would counter the healthcare crisis, partly by promoting preventive medicine in patient care. Ideally, if personalized medicine is implemented in healthcare, this would most possibly lead to disease prevention and, over time, a decrease in healthcare costs. This is mainly because increased investment in personalized medicine would reconstruct the ailing business model of the biotechnology and pharmaceutical industry, decrease the costs of clinical trials and thus lower the prices of drugs, take advantage of a growing IT infrastructure, and promote economic prosperity through increased growth in related manufacturing and industrial sectors and through the growth of medical and scientific centers devoted to the practice of personalized medicine. Furthermore, personalized medicine would lead to a more efficient healthcare system, thus reducing the rate of cost inflation in healthcare and leading to increased economic prosperity and more jobs. Personalized medicine will accomplish this not only through the development of targeted therapies such as Gleevec and personalized genetic profiles, but also through the "role of companion diagnostics which can simplify the drug discovery process, make clinical trials more efficient and informative and be used to individualize the therapy of patients" (Padadopoulos et al., 2006).

At the 2013 ASCO annual meeting, Janet Woodcock of the FDA described a number of challenges of precision medicine, including scientific, policy, logistical, and value-related options for healthcare. Diagnosis is the basis of therapy and personalized medicine, and reliable diagnostics are needed to relay understandable information to clinicians. Next-generation sequencing, as described in Chapter 2, will generate huge amounts of data that will need to be made sense of. As the number of individual biomarkers within a companion diagnostics assay increase, so do the complexity, cost, time, and risk of development. How do we find a balance between finding the "best" set of biomarkers and the practical requirements of companion diagnostic development and commercialization? One option would be to choose the best set of biomarkers that are compatible with the development timelines, or finding the right patients to aim for the best coverage, even if this means slowing drug development and missing a few patients. Gene panels are now being used by diagnostic companies, but they may miss mutations that next-generation sequencing would find.

According to Richard Watt of Qiagen, new technologies are enabling new paradigms in the way we think about disease management and patient care. We seek to expedite ways to safely translate these technologies for diagnostic and therapeutic use indiscriminately. The demand from pharma for regulated companion diagnostic tests and the complexity of the development process have resulted in the formation of many physician–pharma partnerships and partnership models. The central element of these collaborations is nearly always the same—"develop and commercialize an FDA approved companion diagnostic alongside the drug ... (without delay!)." Although the basic framework is very clear, there are still many related questions which are as yet unresolved.

The history of personalized medicine goes back to December 1984, when elite geneticists set out to sequence the entire human genome. On April 14, 2003, this moonshot was achieved. A number of trends have emerged since then, which are detailed in this book. Personalized medicine by the numbers report FDA report 155 drug labels of genomic guided therapy. According to Geoffrey Ginsburg, PhD, of the Duke Center for Genomic and Personalized Medicine, personalized medicine will lead to prenatal sequencing and newborn screening. Ginsburg cites the proliferation of targeted therapies, including

- Vermurafenib (ZEBORAF®) approved for BRAF V600E–positive melanoma
- Trametinib (MEKINIST®) approved for BRAF V600E– and V600K–positive melanoma
- Ado-trastuzumab emtansine (KADCYLA®) approved for HER2-positive metastatic breast cancer

There have also been gains in cancer diagnosis and prognosis:

- Oncotype DX®: 21-gene RNA signature from breast tumor, 12-gene RNA signature from colon tumor

- MammaPrint®: 70-gene RNA signature from breast tumor (FDA approved)
- BluePRint™: 80-gene RNA signature that distinguishes basal, luminal ERBB2 subgroups of breast cancer
- Pathwork® Tissue of Origin test: 2000 RNAs to classify cancer of unknown primary (FDA approved)
- AlloMap®: 11-gene RNA signature for rejection following cardiac transplant (FDA approved)
- Corus™ CAD: 23-gene blood RNA signature for coronary artery disease (CAD) obviates the need for coronary angiography
- Triage® Cardiac Panel: 5-blood-protein signature for assessment of chest pain and shortness of breath

There are other diagnostic tests in the pipeline for lupus, breast cancer, viral infection, pulmonary fibrosis, and Alzheimer's disease.

In 2005, the genetic etiology of age-related macular degeneration (AMD) was revealed through genome-wide association studies: a variant complement factor H responsible for AMD. Fast-forward to today, and thousands of variants have been found for chronic disease and complex traits, a stunning accomplishment and underpinning of disease, even if these results have been accompanied by a disappointing odds ratio of less than 2.

Today, cancers such as lung adenocarcinoma are being diagnosed based on their genetic alterations rather than histologic subtypes, and treated with targeted therapies based on their underlying somatic mutations, as evidenced by Xalkori.

Mobile health and social networks are also undergoing a transformation, with technology playing a large role. Cell phones now have the capacity to delivery important health information. There now exist a shear proliferation of products that can monitor phenotypes and different patterns of disease, such as Google's glucose sensors and contact lenses sensing interocular pressure. Social media will indicate health trends, such as Google Flu Trends and Twitter Data Models. Ginsburg predicts that this summation of data streams will constitute a better and more precise representation of health and disease states through social network data, sensor data, and predictive modeling. He added that community hospitals need to keep up with the technology of personalized medicine and providers and staff in community hospitals lack the requisite education on genomic medicine and genetic testing.

CHALLENGES AND INCENTIVES FOR IMPLEMENTATION

Components of the book

In Chapter 1, "Introduction: Biomedical Innovation and Policy in the Twenty-First Century," I introduce the concept of personalized medicine, presenting such concepts as pharmacogenomics, targeted therapies, and individualized diagnosis and treatment. This chapter introduces the central thesis of the book: personalized medicine will have a major impact on medicine and healthcare, and it will have significant socioeconomic effects on society and will rein in healthcare costs.

Chapter 2, "The Rise of Genomics and Personalized Medicine," deals with the innovations in medicine as a result of genomics and the Human Genome Project. The Human Genome Project, which resulted in the complete sequencing of the human genome by 2003 through public and private ventures, led to a revolution in medicine through the development of genomic technologies. Basically, biomarkers could be discovered in the human genome that would lead to potential disease therapies by diagnosis of disease in patients through companion diagnostics. This chapter concerns the science behind personalized medicine and pharmacogenetics and genomics, which essentially stems from the Human Genome Project. Here, I define terms such as *clinical utility, whole genome sequencing, omics, genome-wide association studies, companion diagnostics*, and *pharmacogenomics* in underscoring the significance of finding the molecular causes of disease.

Chapter 3, "Patient Narratives: Personalized Medicine in the Field," underscores many of the stories of patients and labs personally affected by the advancements in genomic and personalized medicine. Detailing the story of the Beery twins, diagnosed and cured by personalized medicine, and the work of Hunter Labs, a clinical laboratory testing facility in Campbell, California, conducting work at the cutting-edge of personalized medicine, this chapter presents "personalized medicine in the field," ethnographic work revealing many of the insights of the stakeholders with stories of personalized medicine.

Chapter 4, "Alliances: Knowledge Infrastructures and Precision Medicine," is about the new initiative proposed by the National Academy of Sciences, led by clinicians and scientists at major medical centers, including Harvard University and the University of California, San Francisco. Precision medicine is a new form of personalized medicine designed to take advantage of biological information generated from sequencing centers, and channel it into a multipronged pathway that would ultimately enable better health outcomes. Based on a novel taxonomy of disease, precision medicine enables an information commons and knowledge network of disease that would include molecular data encompassing individuals' genomes, transcriptomes, epigenomes, proteomes, microbiomes, metabolomes, and exposomes, incorporated with traditional taxonomies based on signs and symptoms.

Chapter 5, "Great Strides in Precision Medicine: Personalized Oncology and Molecular Diagnostics," is about the tremendous advances in cancer diagnosis and treatment, with new guidelines for the molecular classification of tumors. Trends in personalized oncology are also revealed, such as cancer vaccines, liquid biopsies, and immunotherapies.

Chapter 6, "Personalized Medicine's Impact on Disease," is about how precision medicine has moved beyond oncology, albeit slowly, through genome-wide association studies and pharmacogenomics for diseases such as type II diabetes, cardiovascular disease, and psychiatric diseases. Expert opinions are included in an understanding of how personalized medicine can have an impact on these diseases.

Chapter 7, "The Genome in the Clinic: Diagnosis, Treatment, and Education," is about whole genome sequencing (which I define in Chapter 2) incorporated into the clinic through the use of molecular diagnostic tests. I describe companion diagnostics and the role of using biomarker profiles in diagnosing and treating patients. I offer many examples of molecular diagnostic companies that are promoting the use of these tests to test patients noninvasively for disease and avoid unnecessary treatment. I also discuss potential educational issues in the context of personalized medicine in this chapter.

Chapter 8, "A New Set of Clinical Tools for Physicians," details how, as a result of the personalized medicine revolution, a new set of clinical tools will be available for physicians in the form of molecular diagnostics. An omics profile will soon be at a clinician's disposal for interpreting patient recommendations. An omics profile consists of the patient's genomic and metabolomic information that will give the physician data on the patient's health status. As a result, rather than merely interpreting a patient's clinical history and biochemical tests, physicians will also be able to recommend lifestyle changes to the patient on the basis of genetic predisposition and disease risk.

Chapter 9, "The Regulatory Landscape," reviews the FDA vision for targeted therapies, companion diagnostics, and biomarkers, and how they will be regulated by the agency. Regulatory drug labels are also discussed.

Chapter 10, "Translational Personalized Medicine: Molecular Profiling, Druggable Targets, and Clinical Genomic Medicine," profiles how efforts in basic research are advancing personalized medicine. From the International HapMap Project to genome-wide association studies, the new genetics based on the Human Genome Project have helped to advance personalized medicine, leading to studies in molecular profiling and somatic mutations. This chapter also chronicles the narratives of the Center for Personalized Genetic Medicine in Cambridge, Massachusetts, and the efforts of Translational Software, a company based in Mercer Island, Washington, which is attempting to bring pharmacogenomics information to the clinic.

Chapter 11, "The Economics of Personalized Medicine," includes an overview of the microeconomic and macroeconomic consequences of these changes in biomedicine and the regulatory issues and concerns regarding payer reimbursement that will figure into personalized medicine. Cost-effectiveness, cost utility, and cost–benefit analysis, in addition to health outcomes and other issues in health economics and microeconomics, will also play prominent roles in determining personalized medicine's economic impact. The distinction between "made-to-stock" drugs and "made-to-order" drugs will be highlighted in this chapter, illustrating why Big Pharma is hesitant to pursue made-to-order drugs for economic reasons.

Chapter 12, "Moral and Ethical Issues: Claims, Consequences, and Caveats," explains how personalized medicine, pharmacogenomics, and genetic testing will lead to a whole set of ethical issues wherein patient privacy, genetic discrimination, eugenics, equity, resource availability, and the ethics of targeted therapy will come into play. In this chapter, the bioethics behind personalized medicine and some necessary caveats to its development and integration into society are explored. The societal consequences of personalized medicine are also discussed.

Chapter 13 serves as the conclusion, summarizing advances, implementation, and challenges.

REFERENCES

Dickenson, D. 2013. *Me Medicine vs. We Medicine: Reclaiming Biotechnology for the Common Good*. New York: Columbia University Press.

Dorsey, E.R. et al. 2010. Funding of U.S. biomedical research, 2003–2008. *Journal of the American Medical Association* 303(2):137–143.

Fontes Jardim, D.L., M. Schwaederle, C. Wei, J. Lee, A. Hong, A. Eggermont, R.L. Schilsky, J. Mendelsohn, V. Lazar, and R. Kurzrock. 2015. Impact of a biomarker-based strategy on oncology drug development: A meta-analysis of clinical trials leading to FDA approval. *Journal of the National Cancer Institute* 107:1–11.

Gambacorti-Passerini, C. 2008. Part I: Milestones in personalized medicine: Imatinib. *Lancet Oncology* 9:600.

Heckman-Stoddard, B.M., and J.J. Smith. 2014. Precision medicine clinical trials: Defining new treatment strategies. *Seminars in Oncology Nursing* 30:109–116.

Joyner, M.J., and N. Paneth. 2015. Seven questions for personalized medicine. *Journal of the American Medical Association* 314:999–1000.

Kornberg, R., and S. Bruder. 2011. Opinion: Federal funding for biomedical research saves jobs, creates jobs. *San Jose Mercury News*, June 25, 2011. http://www.mercurynews.com/2011/06/24/opinion-federal-funding-for-biomedical-research-saves-jobs-creates-jobs/

McNeil, C. 2015. NCI-MATCH launch highlights new trial design in precision-medicine era. *Journal of the National Cancer Institute* 107:4–5.

Padadopoulos, N., Kinzler, K.W., and B. Vogelstein. 2006. The role of companion diagnostics in the development and use of mutation-targeted cancer therapies. *Nature Biotechnology* 24(8):985–995.

Piquette-Miller, M., and D.M. Grant. 2007. The art and science of personalized medicine. *Clinical Pharmacology and Therapeutics* 81:311–315.

Ranade, K., Higgs, B.W., March, R., Roskos, L., Jallal, B., and Y. Yao. 2014. Application of translational science to clinical development. In *Genomic Biomarkers for Pharmaceutical Development*, Yihong, Y., Bahija, J., and Koustubh, R., eds., 1–21. San Diego: Academic Press.

Roper, N., Stensland, K.D., Hendricks, R., and M.D. Galsky. 2015. The landscape of precision cancer medicine clinical trials in the United States. *Cancer Treatment Reviews* 41:385–390.

Simoncelli, T. 2013. Paving the Way for Personalized Medicine: FDA's Role in the New Era of Medical Product Development. Rockville, MD: U.S. Food and Drug Administration.

Soda, M. et al. 2007. Identification of the transforming EML4-ALK fusion gene in non-small cell lung cancer. *Nature* 448:561–566.

CHAPTER 2

The rise of genomics and personalized medicine

The era of the blockbuster drug is fading away.

Richard Weinshilboum, MD
Mayo Clinic

HUMAN GENOME PROJECT, SEQUENCING, AND GENOMIC DATA

Studies in genomics, the branch of molecular biology related to the structure, function, evolution, and mapping of genomes (the entire set of hereditary information coded in an organism's DNA), in principle began with the discovery of the structure of DNA by James Watson and Francis Crick in 1953. DNA, short for deoxyribonucleic acid, is the molecule that contains the cell reproduction directions for most known forms of life, including humans. DNA typically takes the familiar coiled or double-helix ladder shape that Watson and Crick determined. Genomics rapidly advanced with the mapping of the human genome through the Human Genome Project (HGP).

The HGP was the private and public venture to map the genome through the efforts of Craig Venter's laboratory in San Diego and the National Human Genome Research Institute under Francis Collins. Begun in 1990, it took $3 billion and 13 years to complete using 330 machines ($225,000 each). The objective of the HGP was to determine the order of nucleotides in the human genome. Nucleotides consist of a string of bases known as adenine, guanine, cytosine, and thymine, and a shorthand version of referring to them, for example, is AGTCTCGATCA. A string of bases is known as a nucleotide sequence. Once a sequence is known, it is analyzed to identify genes located on it.

The impact of the HGP remains huge and profound. It has provided genome organization information that has stimulated all biological research. Researchers at the HGP found 22,000 genes and discerned only 1%–2% of the genome codes for protein. It introduced what are known as high-throughput methodologies, methodologies that provide data at extremely fast rates. The sequence of one genome cost $200 million in 2001, mainly for reagent costs. The HGP also prompted the development of next-generation sequencers (discussed below).

How is the human genome, the complete set of genetic information, organized? A human inherits 3 billion base pairs from each parent. The sperm and egg contain a haploid set of chromosomes; a haploid cell contains 3 billion base pairs (or 6 billion bases), or 23 chromosomes. A diploid cell results containing 6 billion base pairs (twice the DNA content of a haploid cell) and 23 pairs (or 46) chromosomes when sperm and egg unite. There are 23 pairs of chromosomes, including 22 autosomes and sex chromosomes in each diploid cell.

The sex chromosomes consist of the X and Y chromosomes, with females having a pair of X chromosomes and males having one X chromosome and one Y chromosome. Chromosomes range in size from 42 million base pairs (chr22) to 240 million base pairs (chr1).

The HGP has two main goals, according to Yashon and Cummings. The first is to create maps of the human genome, and the second is to find the location of genes in a genome and locate each gene on a map. Prior to the advent of the HGP, genes for simple, monogenic (one gene for one trait) Mendelian disorders were discovered through a method known as positional cloning. This method was used to map genes for cystic fibrosis, neurofibromatosis, Huntington's disease, and dozens of other genetic conditions, and to map most human chromosomes (Yashon and Cummings, 2009).

In positional cloning, markers are identified that show differences in restriction enzyme cutting sites or differences in the number of repeated DNA sequences (such as short tandem repeats). Once these markers are assigned to specific chromosomes, they were used to follow the inheritance of a genetic disorder in pedigrees, and establish linkage between the marker and the mutant allele for that disorder. Although positional cloning identifies one gene at a time, by the late 1980s more than 1500 genes and markers had been assigned to human chromosomes. Some of the genes mapped by positional cloning include Wilm's tumor, amyotrophic lateral sclerosis, myotonic dystrophy, retinoblastoma and familial polyposis (Yashon and Cummings, 2009).

For several decades now, genes and the disorders they coded for have been determined through a type of study called genome-wide association studies (GWAS). GWAS are an examination of many common genetic variants in different individuals to see if any variant is associated with a trait. GWAS typically focus on associations between single-nucleotide polymorphisms (SNPs) and traits like major diseases to find genetic variations associated with a particular disease. These studies normally compare the DNA of two groups of participants: people with the disease (cases) and similar people without (controls). Each person gives a sample of DNA, from which millions of genetic variants are read using SNP arrays. If one type of the variant (one allele) is more frequent in people with the disease, the SNP is said to be "associated" with the disease. The associated SNPs are then considered to mark a region of the human genome that influences the risk of disease. In contrast to methods that specifically test one or a few genetic regions, the GWAS investigate the entire genome. The approach is therefore said to be noncandidate-driven, in contrast to gene-specific candidate-driven studies. GWAS identify SNPs and other variants in DNA that are associated with a disease, but cannot on their own specify which genes are causal (Manolio et al., 2010). Through knowledge of a million common variants, GWAS surveys look for statistical associations of inheritance of a particular variant and getting the disease. GWAS give a clue as to which variant confers risk for a certain phenotype. The first disorder illuminated through GWAS was age-related macular degeneration. Through more GWAS, new regions are being indicated through these statistical relationships. There are now more than 1600 successful GWAS, advancing the study of complex diseases. As GWAS give clues for where to find variants, these studies gave an indication that variants were present in noncoding regions of the genome, that is, regions of DNA that do not code for genes. Insufficient sample sizes have been known to characterize GWAS; that is, the number of people required to be enrolled to come up with a hit is large (Pearson and Manolio 2008). Currently, a GWAS chip often has more than 1 million variants, and 5 million SNPs from Illumina.

Subsequent to the introduction of GWAS, other methods, faster and simpler to use, have been employed to sequence and map the human genome. This history of DNA sequencing includes the Sanger method, discovered by Frederick Sanger in 1972. Introducing the chain terminator method, the Sanger method was extremely costly and little used until the 1980s and 1990s, when it became semi-automated through molecular biology advances that made the relevant reagents inexpensive and widely available. But even with improvements, Sanger sequencing remained a time-consuming and laborious procedure.

The sequencing challenge included the need for technologies that accurately determine sequences of 6 billion base pairs, with low error rates. One error in every 10,000 bases would result in 600,000 incorrect bases in the genome. The genome must also be efficiently sequenced through fast automation. Sequencing must also be affordable, since the costs constrain applications, and improving error rates drive up the costs. Portability, or the ability to transport genomes once sequenced, is also a concern.

However, it was not until the advent of next-generation sequencing (NGS) that the time and money spent in sequencing the human genome and finding genes and their associated disorders considerably decreased. NGS uses massively parallel methods (a lot of processes simultaneously) that generate short reads of nucleotide bases through an amplification method of some kind. It generally has a 0.001 error frequency for Illumina HiSeq2000 and 0.01 error frequency for Ion Torrent PGM. With NGS, one can perform whole genome sequencing, which captures the entire unique DNA sequence of an organism's genome, and whole exome sequencing, which determines sequences of the regions of DNA in the genome that code for genes. Other methods that can be performed with NGS are whole transcriptome sequencing, which determines the mRNA transcripts for which genes code, and ChIP-seq, which determines the chemical changes made to DNA on a genomic level. NGS generates up to 6 billion base pair reads in a single run (with one run taking several days). NGS is also known as being ultradeep: each base is interrogated on many different reads. The cost per base is a fraction of Sanger sequencing. NGS allows a human genome to be sequenced in days versus years for thousands of dollars versus millions of dollars.

NGS is revolutionizing the study of the genome in many layers by adding to our understanding of the analysis of normal human variation, evolution, Mendelian disorders (disorders caused by a single gene), and cancer. NGS involves chopping up the genome into fragments and gluing them to a glass slide. Through cluster growth, molecules are identified at certain positions and thousands of clones of that molecule are sequenced. Through NGS, genome sequencing is now affordable. The price has been falling significantly per year, and currently hovers around $3000–5000 per genome. Meanwhile, the number of genomes sequenced continues to climb, and sequencing is beginning to migrate out of the research laboratory into the clinic.

What happens to the genome once it has been sequenced? What do we do with all that genomic data? The challenge is to translate all that data into information. This is done through a tool known as read alignment. A read is a short-nucleotide sequence of typically 100 base pairs. In read alignment, a read of 100 bases is mapped to the human reference genome (the actual genome sequenced by the HGP). The human genome is approximately 3.2 billion bases long. To understand the remapping complexity of read alignment, take a 16-inch LEGO ("read") and map it to the human genome (a "LEGO read" runs three-fourths of the way around the earth's equator). For read alignment to work, algorithms must be efficient and the alignment must be able to handle nonperfect matches. It is actually these nonperfect matches that lead to the calling of variants. It is the different number of variants we possess that makes us unique. A single-nucleotide variation (SNV) is a one-base difference in a sequence region.

Mapping reads requires base-to-base matching in which one finds out where the read goes and "sticks" on the human genome. Variant calling involves the detection of differences between the data obtained from sequencing an individual and that found in the reference genome. Also of importance are INDELs. An INDEL is an insertion or deletion of a string of bases in a sequence region. INDELs are found through read alignment and lead to variant calling. For example, to find a deletion, the genome is mapped to the reference genome and a gap in the genome represents a deletion.

One type of read alignment that can provide relevant information is somatic variant detection. Somatic variants are DNA changes or differences that arise in a tumor. Read alignment of a person's original DNA and the tumor DNA reveals differences, since the tumor has its own genome. To determine the somatic or tumor-specific variants, we can simply subtract the normal genome variants from the tumor genome variants. A tumor can contain between several hundred and several thousand somatic variants. A somatic mutation is a mutation that is present in the tumor genome but not in the original genome.

At some point, all of this information has to translate into knowledge. Aligning reads and calling variants provides us with information about an individual's genome. To understand what those variations mean and to generate knowledge, we need to provide context. Context can be provided through annotations, such as the tumor profiling described above, and through visualizations. For example, when studying a condition or a disease that affects members of a family, we sequence the "affected" and additional "unaffected" family members' genes to learn more genetic information about pedigrees and family history. There also must be presentation of variant calls and annotations in a visually appealing way. Genes are biologically interactive, and one way to represent this is by pathways, and layering mutations on pathways can improve interpretation and understanding.

ENCODE: MAPPING THE FUNCTIONAL GENOME

According to John et al. (2013):

> Following the success of the Human Genome Project, the National Human Genome Research Institute (NHGRI) of the US National Institutes of Health (NIH) launched the ENCODE (ENCyclopedia Of DNA Elements) Project Consortium in 2003 (ENCODE Project Consortium, 2007), with the ultimate goal of identifying all functional elements represented in the human genome. Functional elements are defined operationally as DNA sequences that reproducibly give rise to specific biochemically detectable activities, such as RNA transcription, transcription factor binding, or chromatin remodeling. The pilot phase of ENCODE targeted 1% (~30 Mb) of the genome. Of this, roughly 15 Mb was selected to represent well-studied, biologically significant loci such as the CFTR locus, the alpha and beta-like globin loci, and the HOXA gene cluster. The remaining 15 Mb was chosen by stratified random sampling methods designed to select regions with varying combinations of gene density and non-coding evolutionary conservation. The results of the ENCODE pilot project were published in a highly cited paper (ENCODE Project Consortium, 2007)

> and provided many biological highlights illustrating both the density and diversity of functional elements in the genome, and the integration of experimental assays of functional elements with evolutionary conservation. The pilot study also brought to light the general transcriptional promiscuity of the genome, and identified many novel transcription start sites (TSSs) and non-protein coding transcripts. Additionally, the study categorized patterns of histone modifications, transcription factor (TF) occupancy, and chromatin accessibility, as diagnostic of regulatory regions and their transcriptional potential (ENCODE Project Consortium, 2007). The study also showed that specific combinations of assays could be used to localize distinct classes of functional elements, such as the segregation of DNaseI hypersensitive sites encoding promoters vs distal enhancers by virtue of histone modification patterns (ENCODE Project Consortium, 2007).

NEXT-GENERATION SEQUENCING AND MULTIGENE APPROACHES

According to a recent report by McKinsey & Company, the clinical adoption of NGS is a hotly debated topic. The company interviewed experts in the field about the topic and elicited the following comments. Costs have been spiraling down due to advanced technology, but the clinical relevance of NGS has been questioned due to lack of therapeutic intervention gleaned from NGS information. Raju Kucherlapati, Professor of Medicine, and Paul C. Cabot, Professor of Genetics, at Harvard Medical School, sees costs coming down rapidly, leading to rapid adoption since separate tests for *EGFR*, *KRAS*, *BRAF*, and *ALK* could add up to $3000. "You could do it all together using next gen for less than $2000." Mike Pellini, president and CEO of Foundation Health, cites that nearly 50% of the next diagnostic tests for cancers in the United States would benefit from a targeted deep-sequencing approach. In three or four years, a majority of these would be done using next-generation sequencing. There are a number of factors that determine the extent of NGS usage in clinical diagnostics: clinical relevance and actionability of sequencing information in the disease cost, quality of sequencing and ease of use, and the regulatory environment (McKinsey & Company, 2013).

PHARMACOGENETICS–PHARMACOGENOMICS

Pharmacogenomics, a critical component of personalized medicine, is the study of the role of genetic inheritance in individual variation to drug response phenotype. To provide context for pharmacogenomics, let us consider the "therapeutic revolution." The therapeutic revolution is demonstrated by the evolution of the pharmaceutical textbook *Goodman and Gilman's The Pharmacological Basis of Therapeutics*. When first published in the 1940s, it listed only a sampling of drugs. Today, more than a thousand high-quality drugs are listed in it. The textbook shows that the field of pharmacology has evolved considerably over the years, and today it bears a stronger foundation for the development of pharmacogenomics.

Pharmacogenomics consists of a series of clinical and scientific goals. Clinical goals include avoiding adverse drug reactions (ADRs), maximizing drug efficacy, and selecting responsive patients. Scientific goals include linking variation in genotype to variation in phenotype, determining mechanisms responsible for that link, and translating that link into a better understanding, as well as prevention and treatment, of disease.

In determining drug–gene pairs, pharmacogenomics has evolved from one gene, to one or a few SNPs, to pharmacokinetic (drug metabolism) and pharmacodynamic (signaling downstream from a target) pathways and haplotypes (a series of genes on a chromosome), to GWAS, to whole genome DNA sequencing.

In recent years, the Food and Drug Administration (FDA) has issued a series of drug–gene pairs through FDA hearings and relabeling warnings, which included the drugs and drug-metabolizing enzymes that served as targets for the drugs. The following is a list of these drug–gene pairs, which require that the enzyme be active for the drug to be appropriately metabolized. Among these enzymes, variations exist among which the drugs can respond.

Thiopurines (anticancer)–*TPMT*
Irinotecan (anticancer)–*UGTIA1*
Warfarin (blood thinner)–*CYP2C9*, *CYP4F2*, and *VKORC1*
Codeine (pain killer)–*CYP2D6*
Tamoxifen (anticancer)–*CYP2D6*
Clopidogrel (inhibits blood clots)–*CYP2C19*

Consider the drug–gene pair thiopurine (an anticancer agent)–*TPMT*. *TPMT* codes for the enzyme thiopurine methyltransferase, which is involved in the metabolism of 6-mercaptopurine. Most individuals are homozygous for the *TPMT^H* allele, having both of them and indicating high levels of the enzyme. Other individuals are heterozygous for the *TPMT* allele, having the genotype *TPMT^H/TPMT^L*. These individuals have moderate levels of the TPMT enzyme. However, a small minority of patients is homozygous for low levels of TPMT and possess the genotype *TPMT^L/TPMT^L*. This small percentage of individuals runs the risk of overdosing with standard drug dosages, and also of experiencing bone marrow failure. Thus, the *TPMT* genetic polymorphism has clinical consequences such that those individuals with low TPMT have a higher risk of thiopurine toxicity and secondary neoplasm or tumor. A high level of TPMT indicates a decreased therapeutic effect of the drug thiopurine.

Another example of pharmacogenomics is the drug Herceptin. Patients that are positive for *HER2* (human epidermal growth factor receptor) are responders to the drug Herceptin. The *HER2* gene, which encodes the growth factor receptor HER2, is amplified and HER2 is overexpressed in 25%–30% of breast cancers, increasing the aggressiveness of the tumor. Slamon et al. (2001) evaluated the efficacy and safety of Herceptin (trastuzumab), a recombinant monoclonal antibody against HER2, in women with metastatic breast cancer that overexpressed *HER2*. The addition of Herceptin to chemotherapy resulted in longer survival times. Thus, according to the study authors, Herceptin increases the clinical benefit of first-line chemotherapy in metastatic breast cancer that overexpresses *HER2* (Slamon et al., 2001).

The clinical implementation of pharmacogenomics involves clinical utility; that is, do pharmacogenomics tests allow for a decision to be made in order to treat patients? Also, practice guidelines for ensuring that pharmacogenomics meets clinical standards must be put into place. Genotyping and sequencing to be used for the purpose of pharmacogenomics screening should take place in Clinical Laboratory Improvement Amendments (CLIA)-approved laboratories (CLIA cover more than 200,000 laboratories and ensure quality laboratory testing), and the data from these tests should be stored in patient electronic medical records. Pharmacies must participate in the process of pharmacogenomics testing, diagnosis, and treatment, and patients and physicians must be encouraged to accept pharmacogenomics. In short, stakeholder involvement in pharmacogenomics must be complete and thorough; otherwise, its impact on the role of personalized medicine in patient care will be minimal. If welcomed, pharmacogenomics has great potential to revolutionize healthcare, given that it meets its clinical outcomes.

Dan Roden, MD, assistant vice chancellor for personalized medicine and Professor of Medicine and Pharmacology at Vanderbilt University School of Medicine, declares that the notion of pharmacogenetics variation is much more complex than just a listing of drug–gene pairs, citing the example of Plavix.

Associations between certain variants and certain kinds of drug responses [are complex]. In some cases, there are large effects of genetic variants; in some cases, there are small effects of genetic variants. In some cases, there are common genetic variants, and in other cases, there are rare ones. So, for example, the response to a drug called clopidogrel, marketed under the name Plavix, depends on bioactivation of the drug: the drug has to be metabolized into something else to exert its antiplatelet effects. That bioactivation is under the control of one genetic pathway called CYP2C19. There is a little debate as to whether it is under exclusive control of that pathway. There are people who have loss of function alleles, and it is that genetic variant in the CYP2C19 gene that results in production of protein that is nonfunctional. This constitutes 20% of the Caucasian population; it is common. These people have two copies of the abnormal allele. In our experience at Vanderbilt, that is about 2.7% of our patients; 2.7% of our patients have the genotype CYP2C19*2/*2, meaning patients have two copies of the abnormal version. But there exist also *3, *4, *6, and *8 variants. Those are all loss of function variants. [Answering the question] how many genetic variants there are that affect the response to drugs, the superficial answer for this particular drug would be one. But the correct answer is, there's a gene, and there are multiple variants in that gene that might affect response. People who have only one copy of the abnormal allele *1/*2, *1 being the normal allele. There is some data that they will respond to Plavix a little less well than normal people. It makes that gene important for the response, and it makes the variant pretty common.

The other issue is that CYP2C19 doesn't affect only Plavix. It affects many other drugs. Patients who are *2/*2 don't metabolize the antidepressant Selexa [like] normal people. When they don't metabolize it well, they develop higher drug concentrations, and they actually have a better response because Selexa is the active drug. It's not just drug–gene pairs: it is many, many genetic variants. Some of them apply to many drugs. A famous gene, CYP2D6, has a protein product which metabolizes a quarter of the widely used drugs in the world. Seven percent of the Caucasian and African populations have an abnormal copy of the gene; there are literally hundreds of variants in the CYP2D6 gene that can cause loss of function effects (Personal communication, Dan Roden, 2014).

The examples that Roden offers make clear that the concept of pharmacogenomics variation and the numeration of pharmacogenomics variants are far more complicated than would appear. There can be multiple alleles for a gene that have variable effects on drug response. Knowing that all true variant associations might not be predicated on discovery of a single drug–gene pair, loss- or gain-of-function effects within the population might also be difficult to determine. However, considering this complexity, pharmacogenomics will most likely lead to advances in pharmacology and pharmacological therapeutics.

Pharmacogenetics consists of variations in a single gene or small group of related genes that affect the metabolism and clearance of a drug (pharmacokinetics) or the action and effect of the drug (pharmacodynamics). Pharmacogenomics are variations in several genes or the genome that influence drug handling. ADRs consist of any undesirable side effect or toxic reaction that is caused by a drug administration. Pharmacogenetics tests help to reduce ADRs by selecting a specific medication and optimized dose for each individual, based on their SNPs in the driver gene that is involved in the drug metabolism. Variation in these genes could contribute to differences in enzymatic activity, kinetics, substrate specificity, and stability between individuals, and as such, their ability to respond to a drug (phenotype). The response to drugs depends on many processes, such as route of administration, amount absorbed, metabolism of the drug, and speed of elimination.

Therefore, pharmacogenetics targets include any polymorphic gene that encodes for any of the proteins that are involved in these processes. The best-studied pharmacogenetics targets are those involved in drug metabolism. Criteria useful for pharmacogenetics tests in learning about the enzyme are crucial for the drug metabolism. Genetic polymorphism of the enzyme causes change in enzymatic activity, and as such, it has a significant effect on the correlation of the dose with the plasma concentration. The efficacy or toxicity of the drug correlates with a change in the plasma concentration.

The drug possesses a narrow therapeutic index. There is a population of nonresponders. Examples of drug classes that may benefit from pharmacogenetics testing are cancer chemotherapy, immunosuppressants, antidepressants, anticoagulants, antipsychotics, and hypertensives. Drug metabolism includes a bioactive substrate to an active or toxic compound, and inactivates a substrate, increases water solubility, or extends the elimination half-life of an active or toxic metabolite.

Metabolic reaction classes include

- Phase I: Functionalization reaction of the parent compound and converting it to a more polar metabolite by adding or removing a functional group. Examples include oxidation, reduction, and hydrolysis.
- Phase II: Transferases involved in a conjugation with acetyl, amino acids, sulfate, and glucuronyl groups.

There are two approaches to test how an individual will metabolize a medication:

1. Phenotyping: Testing metabolic activity by administrating a safe drug (probe drug) that is metabolized by the same enzyme to the patient, for example, dextromethorphan for assessing CYP2D6 activity.
2. Measuring the enzyme activity in peripheral blood cells, for example, TPMT for detecting individuals with susceptibility to hematological toxicity with azathioprine (AZA).

The genotyping of various gene modifications might be studied in pharmacogenetics, including SNPs, an array of SNPs on one allele (intragene haplotype), a combination of SNPs on two alleles (haplotype), complete gene deletion, or gene duplication. Variants may occur in coding, noncoding, or regulatory regions of a gene. Genotyping is less expensive, less invasive, and more reproducible than phenotyping tests. However, not all variants are currently detectable by genotyping false negatives. Also, there are cases where the same genotype generates different phenotypes.

Pharmacogenetics testing is based on information derived from *in vivo* human studies and therefore is limited. There is one publicly available research tool for pharmacokinetics; it is a reliable resource for current genetic variations contributing to drug metabolism. PharmGKB is updated by the NIH Pharmacogenomics Research Network (PGRN). For example, dosing guidelines for 20 antidepressant drugs, based on CYP2D6 and CYP2C19 variants, can vary from 20% of the usual dose for a poor metabolizer (PM) to 300% of the usual dose for an ultrarapid metabolizer (UM). Drug companies are now developing companion diagnostic and pharmacogenetics testing, alongside developing new drugs.

TPMT is a phase II metabolic enzyme that acts on 6-MP and the prodrug AZA, detoxifying them. These are cytotoxic drugs used for leukemia and rheumatic disease. Table 2.1 displays the dosing for TPMT and azathioprine according to alleles in the population. A lack of sufficient TPMT activity in bone marrow causes hematologic toxicity. Ninety percent of individuals have high enzyme activity of TPMT, 10% intermediate, and 0.3% very low. The trimodal activity of TPMT is the result of its enhanced proteasomal degradation. TPMT activity is inherited as an autosomal codominant trait with polymorphism. Eleven alleles have been identified, including nine SNPs leading to amino acid substitution, one causing stop codon formation, and one destroying the splice site. Three alleles, *TPMT*2*, **3A*, and **3C*, (amino acid substitution) account for 95% of the intermediate to low activity (higher proteolysis). Heterozygous patients having intermediate and homozygous activity with two variant alleles are TPMT deficient.

There are three routes to test for TPMT status: enzymatic activity in blood cells; measuring the concentration of 6-MP in urine, or genotyping; and using polymerase chain reaction (PCR)-based assays to detect three signature mutations in these alleles. With these, 90% of all variants are identified. *TPMT*3A* is the most common allele in Caucasians, while *TPMT*3C* is the most prevalent allele in both the African and Japanese or Chinese population.

In terms of clinical application, patients with low TPMT activity may tolerate a standard dose but are at greater risk of toxicity. Childhood acute lymphoblastic leukemia patients with TPMT deficiency tolerate a full dose of mercaptopurine therapy for a short period (7% of the scheduled period), whereas heterozygous and wild-type patients tolerate full doses for 65%–84% of the scheduled period (2.5 years). A potential valuable role for TPMT status testing is before AZA intravenous administration for Crohn's disease. The dose has to be reduced by 90% for homozygous *TPMT*, while heterozygous patients only need a 20%–30% reduction in dose.

CYP2D6

CYP2D6 is a phase II enzyme metabolizing more than 100 drugs and environmental toxins. More than 80 variants are described for the CYP2D6 gene, which are grouped based on the four phenotypes they can generate: UM (6%), normal or extensive metabolizer (EM) (60%), intermediate metabolizer (IM) (25%), and PM (10%). Of Caucasians, 5% are UM and 5%–10% are PM. Only 1%–3% of African Americans and Asians are PM. CYP2D6 status can be tested for selecting a medication or dose determination of a substrate. One example is codeine (prodrug), which is activated by CYP2D6 to morphine. An EM would need a standard dose, a PM may need a different analgesic agent, and a UM may require lower doses than EM. Tamoxifen (ER antagonist) is another example of a prodrug that is converted to endoxifen by CYP2D6. Women with deficient CYP2D6 do not respond well to tamoxifen and have a higher rate of relapse. Most drugs are inactivated by CYP2D6, for example, antidepressant drug nortriptyline (NT). In this case, PM requires lower doses than EM to produce similar concentrations of active drug. The accelerated rate of metabolism in the UM phenotype requires megadoses of the same drug (2–12 times the standard), but this may have other ADRs because of the toxic metabolite. So, it is safer to choose a different drug for UM.

CYP2C9 is a member of the CYP2C family, and two of the associated substrates are tolbutamide and S-warfarin. There are three phenotypes, PM, IM, and EM, for this gene, representing 12 allelic variants. Two variants of *CYP2C9*2* and *CYP2C9*3* account for most PMs when homozygous. *CYP2C9*2* has been reported in 8%–19% of Caucasians and 1%–4% of African Americans. *CYP2C9*3* is present in 6%–10% of Caucasians, 1.7%–5% of Asians, and 0.5%–1.5% of African Americans. Examples of therapeutic compounds metabolized by CYP2C9 are nonsteroidal anti-inflammatory drugs (NSAIDs) such as naproxen and fluvastatin. Dose adjustment based on phenotype has been published for warfarin and tolbutamide. Warfarin (anticoagulant) is one of the cases that need multigene evaluation for a drug response.

Two of these genes are *CYP2C9* and *VKORC1*. Weekly doses of warfarin may vary 20-fold between individuals.

Pharmacogenomics of warfarin

Factors affecting warfarin's effect are age, body weight, diet, concomitant medications, and *CYP2C9* polymorphism and *VKORC1* variations. Individuals with *CYP2C9*2* and 3 alleles have impaired warfarin metabolism and higher plasma concentration, which may cause life-threatening

Table 2.1 TPMT and Azathioprine

Phenotype (genotype)	Examples of diplotypes	Implications for azathioprine pharmacologic measures	Dosing recommendations for azathioprine	Classification of recommendation
Homozygous wild-type or normal, high activity (two functional 1 alleles)	*1/*1	Lower concentrations of TGN metabolites; higher methylTIMP; this is the normal pattern	Start with normal starting dose (e.g., 2–3 mg/kg/day) and adjust doses of azathioprine based on disease-specific guidelines. Allow 2 weeks to reach steady state after each dose adjustment.	Strong
Heterozygote or intermediate activity (one functional allele, *1, plus one nonfunctional allele, *2, *3A, *3B, *3C, or *4)	*1/*2, *1/*3A, *1/*3B, *1/*3C, *1/*4	Moderate to high concentrations of TGN metabolites; low concentrations of methylTIMP	If disease treatment normally starts at the full dose, consider starting at 30%–70% of target dose (e.g., 1–1.5 mg/kg/day), and titrate based on tolerance. Allow 2–4 weeks to reach steady state after each dose adjustment.	Strong
Homozygous variant, mutant, low, or deficient activity (two nonfunctional alleles, *2, *3A, *3B, *3C, or *4)	*3A/*3A, *2/*3A, *3C/*3A, *3C/*4, *3C/*2, *3A/*4	Extremely high concentrations of TGN metabolites; fatal toxicity possible without dose decrease; no methylTIMP metabolites	Consider alternative agents. If using azathioprine, start with drastically reduced doses (reduce daily dose by 10-fold and dose thrice weekly instead of daily) and adjust doses of azathioprine based on degree of myelosuppression and disease-specific guidelines. Allow 4–6 weeks to reach steady state after each dose adjustment.	Strong

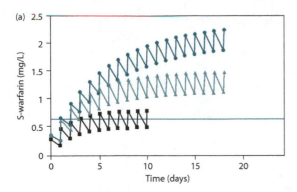

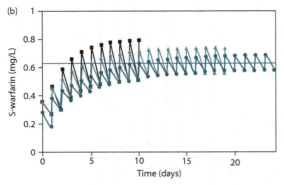

Figure 2.1 (a) Use standard 5 mg/mL. Warfarin for all 3 patient types. (b) Adjust the warfarin dose based on patient's CYPP2C9 allele. CYP2C*1/*1 (black square), which is normal, gets 5 mg/mL; CYP2C9*1/*2 (triangle) receives 3 mg/mL; and CYP2C9*1/*3 (circle) gets 1.5 mg/mL. (Reprinted with permission by Bruns, D.E., Ashwood, E.R., and C.A. Burtis. 2007. *Fundamentals of Molecular Diagnostics*. St. Louis: Saunders, p. 209.)

bleeding. *VKORC1* encodes for the vitamin K reductase complex subunit, also affecting warfarin dosing. *CYP2C9* and *VKORC1* account for 40% of the variability of warfarin's dosage (Figure 2.1).

UDP-glucoronyltransferase

Impairment of this enzyme system leads to the accumulation of metabolite that should be eliminated. There are 117 members in this family, but the most clinically important one is UGT1. UGT1A1 is the hepatic isoform and the most studied one for genetic variations. An important substrate for UGT1A1 is bilirubin, and loss of UGT1A1 activity causes hyperbilirubinemia. More than 30 inactivating SNPs have been identified.

The polymorphic TA repeat in the TATA box of the UGT1A1 promoter may consist of 5, 6, 7, or 8

repeats. (TA)6 (*UGT1A1*1*) is the most common number of repeats (normal allele). (TA)7(*UGT1A1*28*) is the most important genotype for risk of severe drug-induced toxicity. The increase in number of TA repeats may reduce transcription and efficiency and lower enzyme concentrations. (TA)5 and (TA)8 alleles are rare but may also be clinically significant. Irinotecan (CPT-11) is a common chemotherapeutic drug used for the treatment of advanced colorectal and gastric cancer (topoisomerase inhibitor). Irinotecan is a prodrug that is converted to active cytotoxic metabolite SN-38 and finally inactivated by UGT1A1. The risk of cytotoxicity is 12.5% for patients with the TA6/7 genotype and 50% for those with the TA7/7 genotype. Reduced doses for TA7/7 patients are recommended.

Prevalence

(TA)	Caucasians	Asian	African American
6	61%	84%	47%
7	39%	16%	43%

DPYD is involved in the metabolism of 5-fluorouracil, and the mutant allele of DPYD*2A is correlated with severe toxicity and mortality from standard doses of 5-fluorouracil. More guidelines for dose adjustment need to be established when coprescribed medication and environmental factors are involved. Closer integration of pharmacy professionals with clinical labs and clinicians will improve success in pharmacogenetics implementation. Clinical trials for new drugs are considering the benefits of pharmacogenetics testing. Theoretically, it is possible to have an ID card that defines what medication and what dose can be used for a patient based on pharmacogenetics. We are entering pharmacogenomics, which will go well beyond genomics. However, the critical components of pharmacogenomics remain: avoid ADRs, maximize drug efficacy, and prescribe appropriate drugs at the right dosage.

Elaine Lyon, PhD, Professor of Pathology at University of Utah School of Medicine and past president of the Association for Molecular Pathology, sees rewards and risks for pharmacogenomics. It will identify patients most likely to benefit from a particular therapeutic product, those not likely to respond to a therapy, and those likely to have an ADR. However, erroneous results could lead to withholding appropriate therapy or to administering inappropriate therapy. NGS could also lead to incidental findings. Lyon adds that there are reasons to show clinical utility for pharmacogenomics for aiding clinicians in actionability and demonstrating the value of genomic medicine for reimbursement. In the Evaluation of Genomic Applications in Practice and Prevention (EGAPP):

- Insufficient evidence was found to support a recommendation for or against the use of CYP450 testing in adults beginning selective serotonin reuptake inhibitor (SSRI) treatment for nonpsychotic depression.
- In the absence of supporting evidence, the use of CYP450 testing for patients beginning SSRI treatment is discouraged until further clinical trials are conducted.

Pharmacogenetics and HIV, abacavir, and HLA-B*5701

Hypersensitivity to abacavir is strongly associated with the HLA-B*5701 allele. There are near 100% negative and positive predictive values. In 2008, the FDA recommended pretherapy screening for the HLA-B*5701 allele.

EMERGENCE OF PERSONALIZED MEDICINE

According to the National Human Genome Research Institute strategic plan, outlined in 2011, there are five domains of genomics research as it relates to patient care:

1. Understand the structure of genomes
2. Understand the biology of genomes
3. Understand the biology of disease
4. Advance the science of medicine
5. Improve the effectiveness of healthcare

After April 2003, when the HGP was completed, genomic medicine began to emerge as individuals' genomic information became part of their clinical care. While the relevance of genomics to biomedical researchers was apparent when the HGP was completed, genomics has become increasingly relevant to providers in the years since. The path to genomic medicine required many technical advances, including the development of NGS, to fulfill the promise of "base pairs to bedside," or "helix to health."

How variants play a role in determining traits and phenotypes is crucial for understanding the genomic basis for human disease. Out of 3 billion nucleotides in our genome, 3–5 million variants, 150,000 rare variants, 10,000–20,000 structural variants (including deletions and rearrangements), and 20 completely inactivated genes (both alleles, also known as broken genes) have been discovered. The genomic architecture of genetic disease has been elucidated to a certain extent, from rare, simple, monogenic, Mendelian disorders (mutations in a single gene) to common, complex, multigenic, non-Mendelian disease, including hypertension and cardiovascular disease. Constituting the major healthcare burden worldwide, these diseases are caused by a series of variants in which there is a greater contribution of these variants.

When the HGP began in 1990, most of the accomplishments in genomics were in the first domain (understanding the structure of genomes). From 2004 to 2010, most advances in genomics were made in the first two domains (understanding the structure of genomes and understanding the biology of genomes). It is anticipated that from 2013 to 2020, the dominant activities will take place in the second domain (understanding the biology of genomes) and third domain (understanding the biology of disease), which involves getting more information about the genomic basis of disease and understanding the biology of disease. Powerful sequencing technologies have allowed for the understanding of rare diseases and complex diseases of which variants confer diseases. The Cancer Genome Atlas is uncovering the genomic basis of cancer by cataloging genomes of tumors. Looking beyond 2020, the research community anticipates success in the clinically oriented domain with remarkable advances in the discovery of variants responsible for disease. Prenatal and newborn genomic analysis, in other words, sequencing the genome of the expected and newborn baby, will become part of regular diagnostics.

In the context of advances in genomic medicine, what is personalized medicine? One definition according to the Personalized Medicine

Coalition is "products and services that leverage the science of genomics and capitalize the trends toward wellness and consumerism to enable tailored approaches to prevention and care." Genetic variation is what makes each of us unique, but currently genetics is mainly practiced on an epidemiological basis: examining family history is how most clinicians practice genetics.

However, we must ask how important family history is. There are, of course, biases in the reporting of family history, and reporting inaccuracies abound. Genomic sequencing is a constantly improving quantitative version of family history. According to Howard Jacob, PhD, director of the Human and Molecular Genetics Center at the Medical College of Wisconsin, the average SNP has the same impact epidemiologically as salt does on blood pressure. Jacob asks, how do we get away from epidemiologically based parameters to quantitative parameters? His answer: having SNPs and family history together is a better predictor of risk, which allows for quantitative information on top of family history.

Responding to the concept of personalized medicine, physicians maintain that they have always personalized treatment of their patients. While this is true, personalized medicine is about transforming personalized healthcare from an art to a science. It concerns the right treatment for the right person at the right time. We move from one-size-fits-all medicine, medicines used to treat millions, to targeted medicine, medicines that treat smaller numbers of people, treating less through trial and error and more through deduction. Another definition of personalized medicine is tailoring medicine to each patient, classifying individuals into subpopulations that differ in susceptibility to disease and response to specific treatment. Prevention and treatment can be concentrated on those who will benefit, sparing expense and side effects for those who will not. This definition, also known as the President's Council of Advisors on Science and Technology (PCAST) definition, was felt to be too unwieldy. An alternate, revised definition, which appealed more to patients, states that personalized medicine is an evolving field of medicine that uses molecular diagnostic tools to identify specific biological markers, often genetic, to help assess which medical treatment will be best for each patient.

In addition, there are many alternative monikers for personalized medicine, such as "targeted therapy" and "precision medicine," the latter of which scientists prefer, but patients do not like. "Individualized medicine" is most favorably regarded by patients. "Stratified medicine" is the least popular among scientists and patients.

The Personalized Medicine Coalition conducted a survey of 60 people, asking them, what is personalized medicine? Their results: 58 people answered that they did not know what personalized medicine was, and of the two people who answered, they did not really know either. Thus, there exists the daunting task of educating the public on what personalized medicine is, without overpromising what it can deliver. Nevertheless, patients, once informed about personalized medicine, tend to be excited about the opportunities it presents them with, and are willing to pay a premium for it.

What is driving personalized medicine? According to Edward Abrahams, PhD, president of the Personalized Medicine Coalition, the prospect of safer, more effective drugs; faster times to cure; and cost-effective healthcare delivery at the bedside are encouraging the implementation of personalized medicine. However, as payers (insurance companies) remind us daily, Abrahams states, "we can't just innovate. We have got to figure out a way to be more efficient—but we don't have the evidence to show that if we implement personalized medicine, it can actually save money." Abrahams cites a recent study by Mount Sinai School of Medicine, which is studying what starting a personalized medicine center does for the economy (see Chapter 9).

The benefits of personalized medicine include the diagnosis of disease more accurately, the selection of optimal treatment and targeted medicines more precisely, an increase in drug safety, the reduction of ADRs, and the transformation from illness reaction to prevention, thus improving healthcare quality, accessibility, and affordability. For example, many major drugs are ineffective for many patients, and obviously to them a waste of money. Furthermore, ineffective treatment can cause harm through adverse side effects. There are an estimated 100,000 deaths per year due to ADRs, the sixth leading cause of death, experienced by 7% of patients, or 2.2 million patients, according to the Personalized Medicine Coalition.

The old paradigm of medicine was based on reactive medical care: the diagnosis came first, and then the drugs were selected with an iterative switching of drugs. The new paradigm of medicine uses diagnostics that determine which drugs may work best based on the molecular profile of the patient.

John Lechleiter, CEO of the drug company Eli Lilly, states that the "power in tailored therapeutics is for us to say more clearly to payers, providers, and patients: this drug is not for everyone, but it is for you." Without big pharmaceutical companies playing a major role, the implementation of personalized medicine will be slow. However, very few companies are not invested in targeted therapies. While the trend is that research and development costs are going up and drug approvals are going down, the good news is that the cost of information is declining. The new wave of the $1000 genome can do much in terms of possibly developing personalized therapies.

Currently, there are 400 personalized drugs approved by the FDA, with more than 100 more in clinical trials. However, we do not know who is going to respond to these drugs; thus, we need diagnostics. Enter companion diagnostics, a diagnostic test that accompanies a therapeutic drug designed to reveal biomarkers for response to a certain drug. A prominent example would be Oncotype DX, manufactured by Genomic Health of California, which reveals drug response biomarkers for breast, colorectal, and prostate cancer.

Companion diagnostics may represent the future of cancer care. The FDA would like to see drug development using biomarkers linked with a companion diagnostic that would be jointly marketed with the drug. This is known as "theranostics," a new term in the personalized medicine era combining therapy and diagnostics. Our knowledge of the genetic mutations underlying cancer is growing exponentially. We are now at the point where it takes four years to go from discovery to targeted treatment.

In addition, many biopharmaceutical companies have launched therapeutics based on companion diagnostic tests (including Roche, Novartis, AstraZeneca, Amgen, and Pfizer). It was projected for 2014 that marketed therapeutics with the associated companion diagnostic test will generate $12 billion in revenue. Personalized medicine depends on investment as well, since investors in the field would like to get a return on their investment. In this respect, there is significant growth for commercial personalized medicine diagnostics, mostly dominated by oncology.

Cloning of the human genome raised great expectations that knowledge of inherited, germline single nucleotide polymorphisms and of tumor-associated somatic genomic abnormalities would lead to better understanding of cancer biology and oncogenesis, with accompanying advances in diagnosis and treatment. Indeed, the mutational spectrum of most cancers has now been defined by The Cancer Genome Atlas (TCGA) project, in which several hundred cases of many of the common human cancers have been fully sequenced, with the results made publically available. These findings have led to a plethora of different types of cancer-directed diagnostic tests, yet it is unclear whether they represent a step forward in patient care or simply an over-hyped fad (Hayes and Schott, 2015).

BIOMARKER DISCOVERY

Biomarkers are objectively measurable indicators of biological states. In healthcare, biomarkers can provide information on the presence and classification of disease, or susceptibility to disease in an individual, or predict or monitor patient response to therapeutic interventions. The evaluation of new biomarkers in cancer medicine is directed toward improving diagnosis or treatment, and thus improving health outcomes and reducing the social and economic impact of the disease. The evaluation of biomarkers has accelerated due to advances in genomics research and associated technologies. Such research has resulted in improvements in our understanding of the genetic basis of cancer, the identification of individual differences in outcomes from disease and response to treatments, and allowed for tailoring of diagnostic testing, treatment, and monitoring of individual patients. Although there are many possible reasons to evaluate biomarkers, including diagnosis and monitoring of disease, much of cancer biomarker research and development is focused on the identification of prognostic and predictive markers. Markers that are associated with survival or other clinical endpoints independently of any specific treatment are classified as prognostic markers. Markers that are associated with clinical benefit but also correlate the effectiveness of a particular treatment are designated predictive markers (also known as effect modifier biomarkers) and thus can be used to select patients for therapy. Certain biomarkers can be both prognostic and predictive. For example, the amplification of human epidermal growth factor receptor-2 (HER2) in breast cancer is associated with poor prognosis with conventional treatment but predicts for benefit from the HER2 monoclonal antibody trastuzumab. Successful clinical validation of a biomarker requires both a good assay and a robust validation strategy. To provide meaningful and interpretable information, assays should: (1) measure what they claim to measure, (2) be reproducible, and (3) produce results that are statistically meaningful. In addition, it is not sufficient to validate biomarkers on unique experimental platforms beyond the means or expertise of a routine clinical molecular diagnostics laboratory.

To be useful, a biomarker assay should be compatible with general clinical laboratory practices. In this phase of biomarker development, the research grade assay's analytical performance characteristics are evaluated and optimized so that they can be tested in samples that reflect the targeted population. To show clinical validity, the assay must characterize the biomarkers in specimens collected in routine clinical practice from patients within the clinical context and intended use of the biomarker. These results can then be translated into clinical sensitivity and specificity. The process of gathering evidence to support clinical utility begins before a test is introduced into clinical practice, and continues following its clinical uptake. It is important to note that regulatory agencies with oversight for review and approval of devices such as the FDA, and certification of clinical laboratories such as Clinical Laboratory Improvements Amendments (CLIA), do not necessarily require evidence of clinical utility for their evaluations of clinical tests. FDA review of a biomarker test has focused principally on analytical and clinical validity, but not on demonstration of clinical utility (Dancey, 2014).

Thus, we have come a long way from the origins of the HGP, and identifying genes through laborious methods. The $1000 genome (performing whole genome sequencing on the genome) may very well be the future, leading to a whole host of biomarker discoveries and genetic tests. Gene panels are being used to diagnose driver mutations. Pharmacogenomics has great potential to affect the way drugs are prescribed, leading to less ADRs and better efficacy of drugs. Personalized drug therapy in its most sophisticated form uses biological indicators, or "biomarkers"—such as variants of DNA sequences, the levels of certain enzymes, or the presence or absence of drug receptors—as an indicator of how patients should be treated, and to estimate the likelihood that the intervention will be effective or elicit dangerous side effects. Companion diagnostics and its associated theranostics also have great potential to improve patient care. Supported with investment by the pharmaceutical industry, personalized medicine is anticipated to play a major role in healthcare, improving patient outcomes and possibly reducing costs.

REFERENCES

Bruns, D.E., Ashwood, E.R., and C.A. Burtis. 2007. *Fundamentals of Molecular Diagnostics*. St. Louis: Saunders.

Dancey, J.E. 2014. Biomarker discovery and development through genomics. In *Cancer Genomics: From Bench to Personalized Medicine*, Graham, D., Berman, J.N., and Arceci, R.J., eds., 93–107. 1st ed. Waltham, MA: Academic Press.

ENCODE Project Consortium. 2007. Identification and analysis of functional elements in 1% of the human genome by the ENCODE pilot project. *Nature*, June 13, 2007.

Hayes, D.F., and A.F. Schott. 2015. Personalized medicine: Genomics trials in oncology. *Transactions of the American Clinical and Climatological Association* 126:133–143.

John, S., Wang, H., and J.A. Stamatoyannopoulos. 2013. Mapping the functional genome: The ENCODE and Roadmap Epigenomics projects. In *Genomic and Personalized Medicine*, Geoffrey, G., and Huntington, W., eds., 28–40. 2nd ed. Waltham, MA: Academic Press.

Manolio, T.A., Guttmacher, A.E., and T.A. Pearson. 2010. Genomewide association studies and assessment of the risk of disease. *New England Journal of Medicine* 363(2):166–176.

McKinsey & Company. 2013. *Personalized medicine: The path forward*. Washington, DC: McKinsey & Company. Available from http://www.mckinsey.com/~/media/McKinsey/dotcom/client_service/Pharma%20and%20Medical%20Products/PMP%20NEW/PDFs/McKinsey%20on%20Personalized%20Medicine%20March%202013.ashx.

Pearson, T.A., and T.A. Manolio. 2008. How to interpret a genome-wide association study. *Journal of the American Medical Association* 299(11):1335–1344.

Slamon, D.J. et al. 2001. Use of chemotherapy plus a monoclonal antibody against HER2 for metastatic breast cancer that overexpresses HER2. *New England Journal of Medicine* 344(11):783–792.

Yashon, R., and M. Cummings. 2009. *Human Genetics and Society*. Belmont, CA: Brooks/Cole.

CHAPTER 3

Patient narratives: Personalized medicine in the field

What is the precise impact of personalized medicine on patients and laboratories worldwide that are conducting tests to diagnose, prevent, and treat illness? The actress Angelina Jolie is a case in point. After learning from a genetic test that she possessed the BRCA gene, which increases her chances of acquiring breast cancer, and considering her family health history of breast and ovarian cancer (Jolie's mother had died of ovarian cancer), Jolie underwent a double mastectomy. Jolie's case sparked a news flurry in the media and the blogosphere on the ethics of personalized genetic testing and its consequences, and exemplifies the debate that will (and should) ensue when personalized medicine enters the clinic.

This book underscores the need for personalized medicine to be implemented into the clinic. According to Dan Roden:

> It is important to understand what we mean when we say personalized medicine. Many people say personalized medicine is the equivalent of genomic medicine. I think personalized medicine is a concept that is much broader than genetics: it has to do with understanding who your patient is, their ability to pay for their medications, and [concerns] like their understanding of the need for certain kinds of medications. Those kinds of things contribute to what an individual physician might do to tailor treatment to a patient's particular requirements. You can take personalized medicine even further than that. You can say that the idea that when somebody enters the hospital, you would like to prevent bedsores. There are patients at high risk for developing bedsores. Personalized medicine has always relied on an individual physician's ability to perceive the need to tailor treatment for a particular patient. [For example], a 95-year-old frail women admitted from the nursing home to hospital would be at high risk for developing a bedsore. You can do specific things to prevent that. So that's personalizing medicine just as much as saying that a person has a genetic variant and thus prescribing a lower- and higher-dose amount of drug.
>
> When people talk about personalized medicine these days, what they mean is the application of genomic or other kinds of marker discoveries that people have developed over the last decade or two that allow us to identify individuals who on the surface are like everyone else but whose makeup makes them more or less susceptible to have good or bad responses to certain drugs. [Even if] all of medicine is personalized since the time of Hippocrates, [one can still] focus on the newer, mechanistic [applications.]

There are many stories of individuals personally affected by the "newer, mechanistic applications" of personalized medicine; two of them are recounted here. One relates to the Beery twins, who went through a long, arduous medical journey ultimately to be diagnosed and cured through a blood draw used for whole genome sequencing. The other is that of Hunter Labs in Campbell, California, whose visionary leader, Chris Riedel, is at the forefront of advancing the beneficial impact of personalized medicine on patients through biomarker testing. They are examples of "personalized medicine in the field."

Personalized medicine has the potential to have great positive impact on patient care, as reflected in the story of Alexis and Noah Beery, twin daughter and son of Joseph and Retta Beery, born on August 16, 1996. Retta Beery articulately recounts the profound story of the discovery, through the tools of personalized medicine, of a mutation that the twins possessed and a journey that was filled with trials and tribulations, which their parents went through to diagnose the twins' unknown disorder. Alexis and Noah Beery were born with developmental disabilities of unknown cause, and the Beerys found themselves "quickly in a new world of specialists, doctor's appointments, and hospital visits." Gastrointestinal testing on Noah and Alexis revealed nothing. The twins were then put on a waiting list for physical therapy, occupational therapy, and early intervention therapy. Their pediatric neurologist ordered an MRI on Alexis, and the MRI came back showing nothing of significance. Metabolic tests showed that the twins did not have metabolic disorder. Retta Beery notes that "this is what people go through every day: that huge amount of testing to figure out what you're dealing with." For the next couple of years, Noah underwent an endoscopy to find out why he was throwing up, and two nuclear tests were performed where Noah sat in a machine for four hours to see if he was digesting food. In the next three years, three more MRIs were done on Alexis. The MRI revealed damage in the ventricular area of her brain, termed periventricular leukomalasia, which is under the category of cerebral palsy. Thus, the twins were diagnosed with cerebral palsy under the assumption that they had lost oxygen while in the womb. The MRI that revealed the brain damage was also done on Noah, and Retta got confirmation of this diagnosis from "books (she) read from doctors from Johns Hopkins."

The Beerys moved forward with the diagnosis of cerebral palsy and based all the twins' therapies around it. At one year of age, Alexis had her first grand mal seizure and was taken to the hospital. Alexis had other symptoms: her eyes would roll up into her head for hours at a time, her arms would be glued to her chest, her hands would point down, and her body would tremor. Alexis's neurologist started her on seizure medication at one year of age. Alexis's body temperatures would also spike very high. She was taken to an urgent care clinic where her temperature read 107.5°. An ambulance was called several times to take her to the hospital, and it was found that urine was refluxing up into her kidney.

Retta Beery notes that there were many difficult and invasive tests. Both twins were undergoing physical therapy, occupational therapy, and speech therapy. Noah and Alexis both had the same neurologist; both had a gastroenterologist; and Alexis had a urologist, orthopedic surgeon, pediatric ophthalmologist, and allergist. Meanwhile, when the twins were age 5½, during Retta Berry's extensive research, she came across an LA Times article that spoke of a disorder that mimics cerebral palsy but can be treated with medication. Mrs Beery found that the common thread that separates the disorder from cerebral palsy is that the patient functions at a higher level in the morning, and as the day goes on, he or she becomes more debilitated, which Mrs. Beery observed in the twins. "When I read the article, I knew this was what we were dealing with. We shot an email off to a neurologist out of the University of Michigan, Dr John Fink, on April 10, 2002." Not long afterwards, the Beerys flew to Michigan to see Dr Fink. Dr Fink examined Alexis and Noah with all the scans and tests that Retta Beery brought. Dr Fink thought there was the possibility that the twins had dopa-responsive dystonia, the subject of the LA Times article. At Michigan, the Beerys took Alexis to a restaurant and Alexis could not get out of the car; they had to carry her to the restaurant, while Retta held her up. Alexis could not swallow. In the meantime, she had lost 25% of her weight. Now, the Beerys were starting to think about feeding tubes and wheelchairs.

At the clinic, Dr Fink prescribed L-Dopa for Alexis. Alexis responded immediately. She slept through the night for the first time ever. The next morning, she woke up and actually walked to the rental car and got into it on her own. Alexis had an immediate dramatic response to the medication. The Beerys saw her do things she had never done before. A couple of months later, Dr Fink confirmed that Noah also had dystonia, and he too responded positively to L-Dopa. Retta adds, "We had a completely different outcome of life based on an article that I had found in 2002. Alexis and Noah started playing basketball and soccer and doing all kinds of things."

Based on her experiences, Retta started a website in 2003 through which she reached people around the world. For people that had dealt with misdiagnosis, the Beerys were always asked through the website, "How do I find out if my child has this diagnosis?" But there was no diagnostic tool available—only a response to the medication. In 2008, Invitrogen, a life science company based in Carlsbad, California, was trying to recruit Joseph Beery to be company CIO. Invitrogen was getting ready to buy another life science firm, Applied Biosystems, and needed Joe's help to oversee the two companies' merger. At dinner with a director of Invitrogen, the Beerys shared Alexis and Noah's story with him, and the director told them that Applied Biosystems made instruments that could diagnose children like Alexis and Noah at birth. That night, due to the conversation they had, it became clear to them that Joe had to take this offer from the life sciences company.

In the meantime, the Beerys were dealing with another medical issue concerning the twins. Alexis frequently suffered from a severe cough and difficulty breathing, so much so that she would turn blue; she went to the emergency room seven times in two months. Retta notes, "We were back into the medical odyssey again for 18 months, with multiple pulmonologists, running multiple tests, and still no one could figure what was going on." To keep her vocal cords open, Alexis received racemic epinephrine, a drug only available through a hospital pharmacy. "It's not something you want your patient inhaling every day." Mrs Beery felt that this cough was related to a neurological issue, but they had exhausted all conventional tests after 18 months.

Then came another pivotal moment, at a conference where Eric Topol, currently director of the Scripps Translational Science Institute in La Jolla, California, was speaking and talking about sequencing and advances that have been made in the discovery of mutations, which were being found every month. Retta asked researchers at the conference if there was any way "we could pay to have Alexis and Noah's whole genome sequenced." After a series of emails through the Beerys' connections at Life Technologies, of which Invitrogen was a subsidiary, Jim Lupski, from the Baylor College of Medicine, heard of Alexis and Noah's story through a PBS special and suggested sequencing the Beery twins to see if a genetic mutation responsible for their neurological disorder could be found. Retta emphatically noted that "(sequencing) just required a blood draw." Three months later, the Beerys met with scientists and physicians at Texas Children's Hospital. A mutation was found in Alexis and Noah, which they inherited from their parents in the *spr* gene, or sepiapterin reductase gene. A defect in this gene makes one dopamine and serotonin deficient.

The Beerys then began taking Alexis and Noah to neurologist Dr Jennifer Friedman at Rady's Children's Hospital in San Diego. Dr Friedman had experience in dealing with the very few patients who harbored the *spr* mutation and was very familiar with the illness the mutation causes. Dr Friedman suggested adding either a selective serotonin reuptake inhibitor (SSRI) or natural precursor for serotonin, 5-hydroxytryptophan (5-HTP). 5-HTP was brought on slowly to Alexis and Noah's regimen of L-Dopa. After five days, Alexis no longer needed breathing treatment, and after three weeks, she could run track. As for Noah, the Beerys observed improvement in fine motor skills.

Retta stated that "we went through two medical odysseys. Whole genome sequencing was the very first time that we had black-and-white evidence of what (was causing the twins' disorder) and could treat them based on this evidence." Since then, the Beerys have spoken around the world about their story, and their medical journey was part of the 2013 Human Genome Exhibit at the Smithsonian Institute. Mrs Beery has testified at the University of California, San Francisco, in front of the President's Commission to Study Bioethical Issues:

I see what neurologists go through for patients that do not fit certain diagnoses, and their checklist is invasive and extensive. There is a huge financial, emotional, and physical cost. My viewpoint is that we have to have personalized medicine advance; we have to have diagnostic tools like whole genome sequencing available, especially when dealing with unknowns. It was a blood draw in comparison to countless invasive tests. If we had these diagnostic tests available, we would have a lot of cases with people in wheelchairs unable to function who would have a completely different outlook.

The Beerys continue to push forward in patient advocacy by sharing Alexis and Noah's story.

Doctors and treating physicians need this information. [We have] several reasons that we share their story. The thought that people that have a disorder that is treatable is maddening. We give hope to people who are exhausted and have run every test that doctors could think of. The system is broken and we need to change it. I am a huge proponent of whole genome sequencing. Five and ten years from now we'll definitely have that in the clinical setting. From birth we will have our whole genomes sequenced.

In answering how better to implement personalized medicine in the clinic, Beery added that the first step is looking at individual better record keeping and having a family history electronic medical record that is effective and efficient. "Electronic records are vital and need to be implemented across the board. Doctors are told they need to spend less time with patients due to costs, and doctors are frustrated with this situation. If doctors were spending enough time up front with their patients, and we had whole genome sequencing, we would get better diagnostics and ultimately less time spent when we got the diagnosis."

The Beery story is not the sole one recounting the positive effects of personalized medicine. Personalized medicine's main impact will be on better patient care, as also shown by the example of Hunter Labs, a laboratory devoted to personalized medicine tests in Campbell, California. Chris Riedel, CEO of Hunter Labs, notes many instances where his lab has demonstrated personalized medicine's profound influence on patient care.

Riedel states, "Doctors are uncomfortable ordering genetic tests because patients cannot change their genetics. Let me give you the other side of that coin. If you have the apoEE4 allele, you have a 35% higher risk of developing early Alzheimer's disease. But if you take a statin at an early age, you will not develop the disease." (While Riedel remains excited about the hope of personalized medicine, his claims about taking statins and Alzheimer's disease are unsubstantiated and cannot be corroborated).

Who wouldn't want to know that? We also test for factor V Leiden, which is a gene carried by 5%–10% of the population. If you have it, you are susceptible to pulmonary embolism, which kills. When you travel or when you have surgery, you should take precautions. If you don't know you have the gene, how do you know to take the precautions? Seventy percent of pulmonary embolism patients that surgeons see have this gene. For personalized medicine, knowing your genetic makeup, is just amazing.

Riedel's statements demonstrate the emerging potential of personalized medicine; however, experts cannot corroborate them with evidence.

Often, if you know your risks, you can deal with it. However, even if something is not actionable, still if you know you have it, you might change your lifestyle. The poster child now is Angelina Jolie. I am now sure more and more tests that are predictive are down the pike. We are on the cutting edge and have developed some of these tests ourselves. One of the tests we have is for cervical cancer; 99% of cervical cancer is caused by the human papilloma virus. The current test that is out there is miserable; it tells you if you have the DNA, which means you do have the STD, but 90% of the time the disease will cure itself. This test is done with a Pap smear. So if you have an indeterminate Pap smear and test positive for the DNA, are you now going to undergo a biopsy and scar the cervix, and on occasion render the woman infertile? We have developed a test where we are looking for the onco-genes that give early indication of cervi-cal cancer. If that test is negative, it is time to go home! If that test is positive, you need to then take some invasive action. It's wonderful. We have actually developed another test we have just released. Once you have the gene, you make proteins, and we have developed tests for these proteins so we are one step closer to the disease. It is amazing. The problem, though, in personalized medicine is that insurance carriers don't want to pay for it. They don't want to pay for the test. They think that insur-ance carriers will change from patient to patient over time. It's totally misguided.

The factor V Leiden test: everybody should have that. You have probably heard about all these NFL players com-ing down with dementia. Well, some-thing like 70%–80% of these NFL players who have come down with dementia have the apoEE4 allele, the same gene that confers a high risk of Alzheimer's. One thought is to test all kids. If they have that apoEE4 allele, the kids don't play contact sports because it is going to have an effect. Now that we know what is going to happen 20, 40, 50 years later, is it worth it? The parents and child can have an informed decision. [Again, physician-scientists cannot substantiate these statements.]

There is another genetic test called lipoprotein (a). One of five people has it. If you have this gene, you are sus-ceptible to early cardiovascular disease, which is the big killer. It's not affected by the normal things used to prevent or treat the disease. Statin medication is the drug of choice, or a healthy diet or exercise, you can do all those things, and you are still likely to get early car-diovascular disease. The only thing that will touch this gene is a drug called Niaspan, which is a superdose of vita-min B. Now shouldn't everybody know that? Our medical director found he

had it and has a family history of it; now he has his whole family on Niaspan.

Cardiovascular disease is the num-ber one killer worldwide, and today it is almost always preventable or reversible, if you know the risks that you carry. And we have newer tests that will stratify the risks, and [if you are willing to fol-low your doctor's advice] and your doc-tor knows what to do. We have made it easy for them.

Riedel showed a patient profile of tests done through the Hunter Heart Panel. The Hunter Health Panel tests for a number of genetic markers above and beyond the lipid profile normally done in labs and ordered by doctors. (The lipid profile has been the standard of care for 30 years.)

Looking for a risk of heart disease and looking to manage it, this is what you run. Looking at this patient, a 50-year-old male, right in the area where people have heart attacks. From the normal lipid profile, he looks as healthy as a horse. If you see this guy, you are going to tell him to go home and keep doing what you are doing. [However, the additional tests done by the Hunter Heart Panel are actually more indicative of cardiovascu-lar risk than a normal lipid profile.] This is what the advanced risk markers tell you. This guy goes from looking very healthy to having an event [myocardial infarction] just about to occur. One test measures rupture-prone plaque in the arteries. That's what causes heart attack and stroke. The C-reactive protein test mea-sures inflammation in the body. We have learned in the last five to seven years that heart disease isn't as much a lipid disor-der as an inflammatory disease. If you have inflammation, you have events. If you don't have inflammation, you don't have events of myocardial infarction. Because these two tests are abnormally high, the patient has an 11-fold risk of stroke and heart attack. He also has the lipoprotein (a) test. The good news is that you can reduce the amount of rupture-prone plaque in 90 days with the right medications. So if you know what the risks are, you can prevent disease. To me, this is preventive medicine. Right now, the American Pediatric Society is now recommending that all kids be tested at around age 9, because it is a slow disease that you don't feel coming. The first time usually when a patient or doctor knows the patient has an issue is when he has a heart attack or stroke. Then things are bad.

What we really invested in is what to do when you find out about these risks. So we had three of the foremost experts write an algorithm where we first address the inflammation, and the second priority being lower the risks. We are trying to put the expertise of an expert cardiologist in the hands of a primary care physician. That's where you want to prevent the disease, long before it gets to a cardiologist.

However, the insurance carriers only care about the bottom line.

Riedel feels that Obamacare only pays lip service to personalized medicine. His idea of preventive medicine is the lipid profile normally done on patients.

As a politician, he can say for the first time we are paying for preventive care, but the level of preventive care being paid for is miniscule. These tests are going to be paid for by people with good insurance plans or those who go to a new breed of doctor, the concierge doctor. The concierge doctor idea has come up in the last six years. You pay a fixed fee a month and you have immediate access to my doctor. You have a two-tier coverage. For those who can afford to pay for this stuff, they will get this kind of coverage. But for the vast majority of patients, they will not. Now Medicare does pay for all these tests, except the apoE test. The CMS [Centers for Medicare & Medicaid Services] hired a payer to figure out how to pay for these molecular tests. The payer determined that apoE is not a meaningful test because it won't change the course of therapy. They're wrong! They are just wrong! But as regulated industry, what are you going to do about it? What you can do is file a lawsuit with an administrative law judge and the doctor presents the case for the test being needed and the government presents the case for why the test is worthless. The judge makes his decision and 80% of the time he decides against the doctor.

Riedel states that he has not tried this with apoE because it is expensive, but he is thinking about it. "Somebody needs to do it," he adds.

We are so much into prevention it is hard to tell the impact of personalized medicine. We have to measure people over time and we haven't had enough time to do that. There are lots of studies that describe on a test-by-test basis.

I love doing this kind of stuff. I feel like a missionary. My closest friends in life, one of them had a stroke 15 years ago, can't talk. The other had a massive MI three years ago. Neither one of them had the test because it wasn't available. Both of them would be fine today if they had the test and had taken the treatment. So it is very personal for me.

Personalized medicine will ultimately decrease costs, and they will be on Medicare longer because they will live longer. We have seen a dramatic decline in risk, but we need a couple of years more to see if it will reduce costs. It should because it's preventive.

However, Riedel laments that personalized medicine testing is not accessible to all, and that currently only the wealthy or well insured can afford the tests offered by Hunter Labs.

Riedel underscores some of the imperatives of personalized medicine, such as preventive care against heart disease, stroke, and dementia; the decrease in healthcare costs; and longer longevity. Assessing the medical, scientific, and socioeconomic impact of this new and emerging area of medicine will require showing that individual healthcare will dramatically shift from a disease-centered paradigm to a prevention-centered one. Moral and ethical issues surrounding the implementation of new biomedical technologies will also shape society, including concerns about privacy and genetic discrimination. In short, a renaissance in biomedicine will occur, tightly linked to a revolution in healthcare in the years approaching 2020. Personalized medicine constitutes many of the current trends in biomedicine and necessitates the implementation of policies that would lead to innovation in the biomedical sciences and the integration of scientific and clinical processes promoting patient cures through personalized medicine, and making personalized medicine affordable for all. Yet, in closing, Riedel, undoubtedly, and justifiably so, remains very excited about the potential of personalized medicine and genetic tests to diagnose and treat disease. However, many of his claims cannot be corroborated or substantiated by experts.

CHAPTER 4

Alliances: Knowledge infrastructures and precision medicine

PRECISION MEDICINE: ITS POTENTIAL DEVELOPMENT AND IMPLEMENTATION

Questions concerning the capability of information technologies to keep pace with the growth and adoption of personalized medicine have emerged with the enormous amount of molecular data yielded by genomics. The Center for Genomics and Personalized Medicine (CGPM) at the Stanford University School of Medicine is directing a study called GenePool, according to the center's director, Michael Snyder. Through GenePool, patients are recruited to have their DNA sequenced, which is then placed in a giant electronic database. Here, individuals are serving as both research subjects and clinical patients. According to Snyder, "electronic health records need to reorganize in a fashion that would make them more accessible," and there is a huge and fundamental task in "getting data organized properly so that it can be mined, managed, and used for clinical care and research." The center's main occupation is to sequence an individual's genome for the lowest cost possible (currently $3000–$5000), and analyze it for clinical purposes. The result is, with all of that sequencing, vast amounts of information will need repositories to store them in the form of a huge information technology (IT) infrastructure.

Enter precision medicine, a new form of personalized medicine designed to take advantage of biological information generated from efforts such as the CGPM at Stanford, and channeled into

a multipronged pathway that would ultimately enable better health outcomes. Precision medicine is the brainchild of the National Academy of Sciences and was developed by the Committee on a Framework for Developing a New Taxonomy of Disease, which produced the groundbreaking document "Toward Precision Medicine: Building a Knowledge Network for Biomedical Research and a New Taxonomy of Disease." The motivation for the committee's efforts are as follows:

> The Committee's charge was to explore the feasibility and need for "a New Taxonomy of human disease based on molecular biology" and to develop a potential framework for creating one. Clearly, the motivation for this study is the explosion of molecular data on humans, particularly those associated with individual patients, and the sense that there are large, as-yet untapped opportunities to use these data to improve health outcomes. The Committee agreed with this perspective and, indeed, came to see the challenge of developing a New Taxonomy of Disease as just one element, albeit an important one, in a truly historic set of health-related challenges and opportunities associated with the rise of data-intensive biology and rapidly expanding knowledge of the mechanisms of fundamental biological processes. Hence, many of the implications of the Committee's findings and

recommendations ramify far beyond the science of disease classification and have substantial implications for nearly all stakeholders in the vast enterprise of biomedical research and patient care (Committee, 2011, 11).

According to the authors of the document (among whom are a number of academic physicians, scientists, and ethicists from major medical centers, including the University of California, San Francisco, Sloan-Kettering Memorial Cancer Center, and Harvard University), the scope of this project seems astounding, as it consists of integrating biological data with medical and health histories to create a new taxonomy of disease that would be stored in an "information commons" and "knowledge network." According to the document, to realize the full potential of precision medicine, both researchers and clinicians must have full access to the whole repository of biological data linked to patients (Committee, 2011) (Figure 4.1).

Taxonomy consists of the practice and science of classification, and the basis of the project of precision medicine is to use hierarchies of large datasets to form a knowledge network of disease that would be based on a new taxonomy. This taxonomy would include molecular data encompassing individuals' genomes, transcriptomes, epigenomes, proteomes, microbiomes, metabolomes, and exposomes, incorporated with traditional taxonomies based on signs and symptoms. That is, a patient's DNA, gene expression profile, chemical modifications to his or her DNA, protein signatures, micro-organismal features, metabolic characteristics, and gene–environment interactions would be included in large databases that would network with each other, enabling the researcher and clinician to provide a complete and holistic picture of the patient and the mechanisms underlying his or her disease. Frank Witney, CEO and president of Affymetrix, a Silicon Valley biotech company specializing in the manufacture of gene chips and microarrays, refers to this ambitious

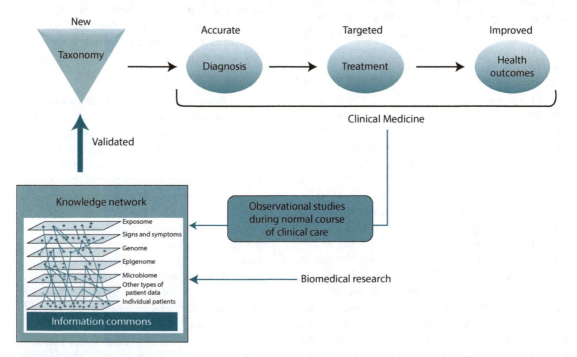

Figure 4.1 Creation of a New Taxonomy first requires an "Information Commons" in which data on large populations of patients become broadly available for research use and a "Knowledge Network" that adds value to these data by highlighting their interconnectedness and integrating them with evolving knowledge of fundamental biological processes. (Reprinted with permission from Committee on a Framework for Developing a New Taxonomy of Disease, National Research Council. 2011. *Toward Precision Medicine: Building a Knowledge Network for Biomedical Research and a New Taxonomy of Disease.* Washington, DC: National Academies Press, 2.)

project as "kind of a moonshot," but says that "all the molecular tools are there."

The committee frames precision medicine as an idea whose time has come, stating in the report that "multiple stakeholders are ready for change," and that public perception toward information and genetic privacy, while skeptical about housing large sets of self-identifying features in databases, is not necessarily unfavorable to the full development of precision medicine. In fact, the authors claim that precision medicine, from where it stands now to its full implementation, would be tantamount to the Internet in the Web's nascence between the mid-1990s to now. In the authors' words, we are at a Flexnerian moment, in reference to the revolutionary report written by Abraham Flexner in 1910 on incorporating professional standards and close ties to research into medicine.

The current taxonomy for biomedicine is the International Classification of Diseases (ICD), based on categories developed by the World Health Organization about 100 years ago. The ICD is now used by physicians to diagnose disease and by insurance companies to reimburse for treatment. The report claims that the current ICD, ICD-10, is outdated, and cannot keep up with an intensive data-driven biology. Future improvements in disease classification must be rooted in the new information commons and knowledge network. The committee argues that constructing and maintaining the information commons and knowledge network will come with its hurdles, but would modernize biomedical research and improve patient care enormously.

How does the committee envision precision medicine to come to fruition? First, the project would be contingent on National Institutes of Health (NIH) funding for the requisite organizations that would enable the production of the information commons and knowledge network. A series of pilot studies of varying complexity would serve as bottom-up efforts to begin the process of developing the information commons. The committee warns that the initiative of precision medicine should not be too rigid, so as to stifle innovation, or too lax, so as to prevent the centralization of efforts required for the development of the new taxonomy.

The new taxonomy promotes a brighter future for research and healthcare, and promises a set of bold consequences that are bound to incentivize it. Among the deliverables are the following:

1. Better health care. The new taxonomy would take advantage of new biological insights to provide more effective targeted therapies for specific patient subgroups, leading to a reduction in wasteful health care expenses, and would lead to more useful characterizations of disease according to molecular causes, which would be catalogued according to disease state more appropriately.

2. Effective utilization of advances in molecular biology, information technology and physiology. With the increasing implementation and use of electronic health records, the IT enterprise is entering the field of medicine. Concomitant with the growth of IT in medicine is the application of molecular biology to medicine that is leading to the "explosion of disease relevant data" and molecularly-informed diseases.

3. A new and improved taxonomic classification of disease. Such a New Taxonomy would
   a. Describe and define diseases based on their innate mechanisms in addition to traditional symptomology.
   b. Go beyond description and be directly linked to a deeper understanding of disease mechanisms, pathogenesis, and treatments.
   c. Be highly dynamic, at least when used as a research tool, continuously incorporating newly emerging disease information (Committee, 2011, 5).

In short, better value in individual-centric healthcare is the result (Figure 4.2).

According to the report, precision medicine is both cost-effective and economically feasible. The current healthcare system, with rising costs (despite advances in medicine and genetics) that are not consistent with healthcare expectations, has questionable value. A "perfect storm" among stakeholders is now on the horizon for the consortium of patients, providers, drug developers, insurance companies, and researchers to take advantage of individual genetic differences that have implications for human physiology. The knowledge network would incorporate information not just from the Human Genome Project, but also from the epigenome, microbiome, and metabolome, in addition to "information on the patient's history of exposure to environmental agents, and psychosocial or behavioral information that would

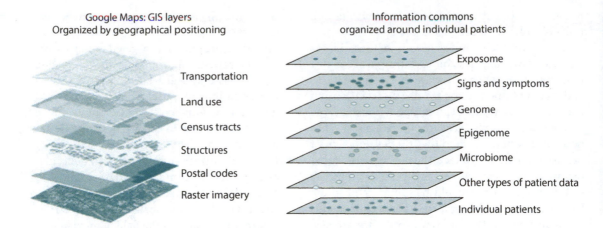

Figure 4.2 The proposed, individual-centric Information Commons (right panel) is somewhat analogous to a layered Geographical Information System (left panel). In both cases, the bottom layer defines the organization of all the overlays. However, in a GIS, any vertical line through the layers connects related snippets of information since all the layers are organized by geographical position. In contrast, data in each of the higher layers of the Information Commons will overlay on the patient layer in complex ways (e.g., patients with similar microbiomes and symptoms may have very different genome sequences). (Reprinted with permission from Committee on a Framework for Developing a New Taxonomy of Disease, National Research Council. 2011. *Toward Precision Medicine: Building a Knowledge Network for Biomedical Research and a New Taxonomy of Disease*. Washington, DC: National Academies Press, 15.)

enhance the diagnosis and treatment of disease" (Committee, 2011, 22).

In short, precision medicine, and its associated information commons, knowledge network, and new taxonomy of disease, are overcoming medical, economic, and technological barriers in order to integrate the biomedical research laboratory, the clinical setting, and healthcare in general. Made mainly possible by advances in IT, molecular biology research data and genome-wide association studies would be routinely used in clinical settings. Society has been ambivalent about the blurring between public and private information through the Internet; however, the notion of precision medicine may obviate concerns about privacy in the information age (Figure 4.3).

IMPLICATIONS OF PRECISION MEDICINE FOR PATIENT CARE AND DISEASE TREATMENT

How will precision medicine affect patient care and treatment of disease? The report states that precision medicine will allow for novel treatment strategies for diseases such as type II diabetes, treatments

that are not in existence currently. Type II diabetes is treated through a traditional signs and symptoms approach, with little consideration given to the molecular causes. By comparison, cancer has been combated with targeted therapy approaches. The report illustrates the differences in current treatment strategies for cancer and type II diabetes. The Committee on a Framework for Developing a New Taxonomy of Disease states:

As illustration, consider the following clinical scenarios; in the first example, molecular understanding of disease has already begun to play an important role in informing treatment decisions, while in the second, it has not:

Patient 1 is consulting with her medical oncologist following breast cancer surgery. Twenty five years ago, the patient's mother had breast cancer, when therapeutic options were few: hormonal suppression or broad-spectrum chemotherapy with significant side effects. Today, Patient 1's physician can suggest a precise regimen of therapeutic options tailored to the molecular characteristics

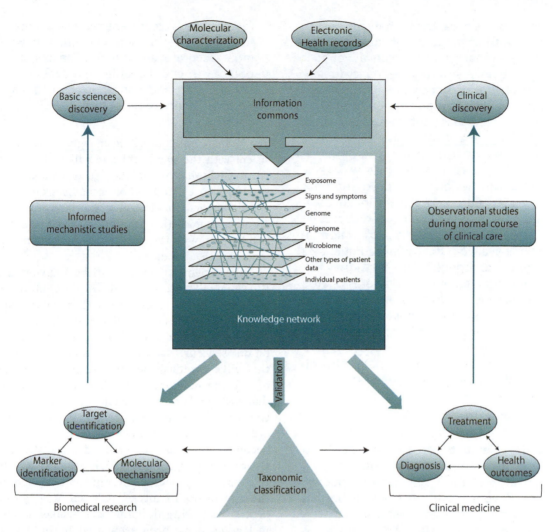

Figure 4.3 Building a Biomedical Knowledge Network for Basic Discovery and Medicine. At the center of a comprehensive biomedical information network is an Information Commons which contains current disease information linked to individual patients and is continuously updated by a wide set of new data emerging though observational studies during the course of normal health care. The data in the Information Commons and Knowledge Network serve three purposes: (1) they provide the basis to generate a dynamic, adaptive system which informs taxonomic classification of disease; (2) they provide the foundation for novel clinical approaches (diagnostics, treatments, strategies); and (3) they provide a resource for basic discovery. Validated findings that emerge from the Knowledge Network, such as those which define new diseases or subtypes of diseases that are clinically relevant (e.g., which have implications for patient prognosis or therapy) would be incorporated into the New Taxonomy to improve diagnosis (i.e., disease classification) and treatment. The fine-grained nature of the taxonomic classification would aid in clinical decision-making by more accurately defining disease. (Reprinted with permission from Committee on a Framework for Developing a New Taxonomy of Disease, National Research Council. 2011. *Toward Precision Medicine: Building a Knowledge Network for Biomedical Research and a New Taxonomy of Disease.* Washington, DC: National Academies Press, 15, 43–44.)

of her cancer, drawn from among multiple therapies that together focus on her particular tumor markers. Moreover, the patient's relatives can undergo testing to assess their individual breast cancer predisposition.

In contrast, Patient 2 has been diagnosed at age 40 with Type II diabetes, an imprecise category that serves primarily to distinguish his disease from diabetes that typically occurs at younger ages (type I) or during pregnancy (gestational). The diagnosis gives little insight into the specific molecular pathophysiology of the disease and its complications; similarly there is little basis for tailoring treatment to a patient's pathophysiology. The patient's internist will likely prescribe metformin, a drug used for over 50 years and still the most common treatment for type II diabetes in the U.S. No concrete molecular information is available to customize Patient 2's therapy to reduce his risk for kidney failure, blindness, or other diabetes-related complications. No tests are available to measure risk of diabetes for his siblings and children. Patient 2 and his family are not yet benefitting from today's explosion of information on the pathophysiology of disease (Committee, 2011, 7).

The authors of the report ask:

> What elements of our research and medical enterprise contribute to making the Patient 1 scenario exceptional, and Patient 2 typical? Could it be that something as fundamental as our current system for classifying diseases is actually inhibiting progress? Today's classification system is based largely on measurable "signs and symptoms," such as a breast mass or elevated blood sugar, together with descriptions of tissues or cells, and often fails to specify molecular pathways that drive disease or represent targets of treatment 2. Consider a world where a diagnosis itself routinely provides insight into a specific pathogenic pathway. Consider a world where clinical information, including molecular features, becomes part of a vast "Knowledge Network of Disease" that would support precise diagnosis and individualized treatment. What if the potential of molecular features shared by seemingly disparate diseases to suggest

radically new treatment regimens were fully realized? In such a world, a new, more accurate and precise "taxonomy of disease" could enable each patient to benefit from and contribute to what is known (Committee, 2011, 8).

The vision seems compelling enough, but is it consistent with the way healthcare will evolve, given technical limitations in the use of databases to store sequences from whole genome sequencing and the capacity of electronic health records? The report maintains that IT is growing at a rate that will keep up with the emergence of the enormous mass of molecular data.

Other scientists, such as Michael Snyder and Rui Chen of Stanford, concur with the authors of the report in their insistence on the revolutionary potential of precision medicine for therapy and treatment. Snyder and Chen write that a revolution is occurring through genomics profiling that will enormously benefit patient care in terms of figuring out disease mechanisms, molecular diagnosis, and personalized medicine. Genomics, transcriptomics, proteomics, and metabolomics are beneficiaries of these technologies and have become "powerful tools for disease studies." Cancer research, for example, is benefitting from this type of approach. Cancer genomes, including those of ovarian cancer, small-cell lung cancer, breast cancer, chronic lymphocytic leukemia, and melanoma, have been sequenced by individual labs or collaborations of labs. Each cancer can be differentiated by different "driver mutations." Cancer genome sequencing can reveal potential targets for disease, as confirmed by a female patient who harbored a p53 mutation, which was responsible for three cancers she developed over five years.

In addition to genomics, the field of metabolomics has proven to be essential in the discovery of a clinical biomarker for personalized medicine. Serum metabolomics reveals many novel metabolic markers of heart failure, including pseudouridine and 2-oxoglutarate, which improve the diagnosis of heart failure (Dunn et al., 2007).

A populations approach to precision medicine can also be useful. The disease risk for one person can be compared with the population risk for the same ethnicity, age, and gender (Figure 4.4).

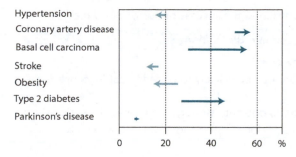

Figure 4.4 Example of a personalized risk graph. Each horizontal line symbolizes genetic risk of one disease tested for a specific individual. The tail of each arrow shows the pretest probability of a disease in a population of certain ethnicity, age, and gender. The front end of each arrow displays the posttest probability with consideration of the person's genomic information. Dark green arrow, increased risk; light green arrow, decreased risk. (Reprinted with permission from Chen, R., and M. Snyder. 2013. *Wiley Interdiscip Rev Syst Biol Med* 5:73–82.)

> In [one study], the genome of a patient was analyzed and increased post-test probability risks for myocardial infarction and coronary artery disease were estimated. Their estimation matched the fact that the patient, although generally healthy, had a family history of vascular disease as well as early sudden death. Genetic variants associated with heart-related morbidities as well as drug response were identified in the patient's genome, the information of which ... may direct the future health care for this particular patient (Chen and Snyder, 2013, 73).

In general, decision support tools have the potential to address these limitations and enable precision medicine approaches to healthcare by providing clinicians and patients with individualized information and preferences, intelligently filtered at the point of care. They will provide clinicians with options for ordering tests: clinicians can indicate the sensitivity, specificity, and positive predictive value of tests. Decision impact tools can also aid clinical workflow by providing algorithms to facilitate decisions on the basis of test results (Mirnezami et al., 2012, 490) (Table 4.1).

MACROECONOMIC CONSIDERATIONS OF PRECISION MEDICINE

The Committee on a Framework for Developing a New Taxonomy of Disease maintains that precision medicine will be cost-effective and pay for itself. Currently, precision medicine has not undergone rigorous economic analysis to ascertain its cost value (such as its cost–benefit ratio) to the healthcare economy. This may be called for shortly; however, on a more transparent level, precision medicine's short-term and long-term impact on the U.S. economy can be preliminarily assessed, if and when precision medicine is implemented. The initial assessment is that through investment in IT and basic scientific research, the impact of precision medicine on the overall economy will be enormous and positive, despite initial costs to generate the knowledge network and information commons. Over time, the healthcare economy will benefit from the decrease in healthcare costs, concurrent with improvements in patient treatment, as a result of the adoption of personalized medicine.

This is partly reflected in many of the recommendations and conclusions of the report. First, with improvements in DNA sequencing technologies, the cost of sequencing a patient's genome will decrease precipitously to relatively affordable prices (the $1000 genome may be in our future), and whole genome sequencing will soon become routine in the clinic. This will rapidly advance personalized medicine through the development and use of targeted pharmacogenomics therapies in patients.

Second, communication between the researcher and the clinician will become less cumbersome and costly. The committee posits that there should be a new discovery model, or a "fundamentally transformed" version of the model. Presently, "when discoveries are judged definitive and potentially useful, an effort is made to return this information to the clinical setting—for example, as a genetic or genomic diagnostic test. This model creates a large gulf between the point of discovery and the point of care with many opportunities for mis- and even non-communication between key stakeholders" (Committee, 2011, 52).

The report states that the current discovery model is unable to take advantage of the explosion of molecular data as a result of personalized

Table 4.1 Healthcare stakeholders and their roles in ensuring the success of personalized medicine

Stakeholder	Recommended actions
Government	Generation of transparent privacy laws
	Identification of socioeconomic priority areas likely to benefit most from precision medicine strategies
	Public consultation regarding "opt-in, opt-out" strategies for research participation
Research industry	Development of effective clinical decision support tools for integration into electronic health records
	Setting up and conducting appropriate pilot studies for data collection in targeted precision medicine areas
Biomedical community	Changes to undergraduate training to develop improved understanding of molecular mechanisms involved in disease
	Development of and contribution to an evolving new system of disease classification incorporating emerging molecular information
	Introduction of a more transparent, participatory role for patients considered for recruitment to clinical trials
Pharmaceutical industry	Development of effective diagnostic tests with or without tandem therapeutic agents for management of conditions identified as major socioeconomic burdens
Patient groups	Increasing participation in health and well-being initiatives
	Use of novel means of providing data for research purposes, including social networks and mobile phone applications
Regulatory bodies	Ensuring that regulatory frameworks are in a place to safeguard patient safety, while ensuring that scientific progress is not hampered

Source: Reprinted with permission from Mirnezami, R., Nicholson, J., and A. Darzi. 2012. New England Journal of Medicine 366:489–491.

Note: Government, industry, patient groups, and regulatory bodies each have a unique role to play in the implementation of precision medicine, from education to safety.

medicine and the integration of personalized "omics" in the clinic. The committee also adds that "perhaps most seriously, the current discovery model offers no path toward economically sustainable integration of data-intensive biology with medicine" (Committee, 2011, 52). They posit a new model that faces the fundamental challenges of making available to the clinician molecular data that would inevitably ease barriers between the researcher and healthcare provider. Great benefits would inevitably ensue, according to the report. While the report remains vague on how this would be implemented, from a preliminary economic standpoint, the adoption of this new discovery model may facilitate cost savings from modifications to the nature of the transmission and exchange of information and ideas between researcher and clinician.

As an example of this novel discovery model, the report mentions a collaborative study between Kaiser Permanente Northern California (Kaiser)

and the University of California, San Francisco (UCSF). For the study, Kaiser patients were asked to contribute their genetic and molecular data and incorporate it into their electronic health records. Despite significant obstacles that mainly emanated from informed consent issues and from the cost of generating the molecular data for Kaiser, the study included well over 200,000 participants, and large-scale generation of molecular data is now under way. The committee report cites this study as an instance of how molecular data can potentially be intimately connected to and directly involved with point-of-care and clinical scenarios.

Third, as socioeconomic factors of illness become included in the knowledge network, their causative aspects in healthcare will be addressed. "Social factors, such as socioeconomic status, quality of housing, neighborhood, social relationships, access to services, and experience of discrimination that can contribute to psychological stress and contribute to poor health and health inequities" (Committee, 2011, 38) are part of what is collectively known as the exposome. The exposome refers to a collection of endogenous and exogenous exposures that affect disease states and are characterized by differential stages throughout an individual's lifetime, from conception to death. A study conducted by researchers at the Harvard School of Public Health concluded that neighborhood socioeconomic status (SES) affects rates of colon and rectal cancer, mediated by certain behavioral risk factors. Their findings suggest that living in a higher-SES neighborhood may protect against rectal cancer in women and colon cancer in higher-educated women (Kim et al., 2010). Similarly, researchers at the University of North Carolina School of Public Health demonstrated that urban stressors contribute to the high incidence of asthma among nonwhites living in Chicago. The study considered "that disadvantaged urban populations experience acute and chronic housing stressors which produce psychological stress and impact health through biological and behavioral pathways," leading to health disparities (Quinn et al., 2010, 688).

Since the knowledge network and information commons would supposedly take into consideration the exposome and thus social factors, the qualified expectation would be alleviation of healthcare disparities and improved socioeconomic conditions for disadvantaged populations. There remains little definitive evidence for this phenomenon taking place at this stage; however, precision medicine creates possibilities for a concomitant lessening of deteriorating health and increasingly better economic situations among certain populations.

Fourth, the resulting enormous investiture in IT and molecular research will stimulate private and public sector elements in the broader economy. Companies such as Google, Microsoft, Apple, and Oracle (Oracle is investing in the development of healthcare databases through its offshoot Oracle Health Sciences) and countless Internet companies and websites will become common terms in healthcare discussions. For example:

Many data sources exist outside of traditional health-care records that could be extremely useful in biomedical research and medical practice. Informal reports from large groups of people (also known as "crowd sourcing"), when properly filtered and refined, can produce data complementary to information from traditional sources. One example is the use of information from the web to detect the spread of disease in a population. In one instance, a system called HealthMap, which crawls about 50,000 websites each hour using a fully automated process, was able to detect an unusual respiratory illness in Veracruz, Mexico, weeks before traditional public-health agencies. It also was able to track the progression and spread of H1N1 on a global scale when no particular public-health agency or health-care resource could produce that kind of a picture (Committee, 2011, 29).

Google has already taken advantage of social networking to mine data for flu trends. The report mentions also briefly the impact and influence of social networking sites that would empower patients and improve social communication regarding health concerns.

Finally, there exists the ultimate criterion that we are potentially getting better healthcare for less cost. Given the huge part of the economy that healthcare represents, we can expect that precision medicine will inevitably impact the overall macroeconomy.

GLOBAL ALLIANCE FOR GENOMICS AND HEALTH

The Committee on a Framework for Developing a New Taxonomy of Disease is not the only consortium launching precision medicine strategies. The Global Alliance for Genomics and Health's mission is to accelerate the progress in human health by helping to establish a common framework of harmonized approaches to enable effective and responsible sharing of genomic and clinical data, and by catalyzing data-sharing projects that drive and demonstrate the value of data sharing. Bartha Knoppers is the chair of the Regulatory and Ethics Working Group at the Global Alliance for Genomics and Health. One of the projects under way at the Global Alliance is the genetic variation project called the BRCA Challenge. The BRCA

Challenge aims to advance understanding of the genetic basis of breast cancer and other cancers by pooling data on BRCA genetic variants from around the world, bringing together information on sequence variation, phenotype, and scientific evidence. Improved understanding of genetic variation in these genes has the potential to improve patient diagnoses and prevention of disease (Global Alliance for Genomics and Health website). Knoppers notes two key studies stratifying subpopulations in 2013 in which large-scale genotyping identified 41 new loci associated with breast cancer risk and a 2014 study citing genetic testing at a sixfold increased risk of prostate cancer. She also cites additional studies urging the need for physicians to understand the contribution of common genetic variations in disease risk, and implications of using genomic information in the risk assessment and risk management of asymptomatic

individuals (Chowdhury et al., 2015), and another in *Science* that copy number variation (duplications) rather than deletions are more likely to be stratified between human populations (Sudmant et al., 2015).

The Global Alliance is aware of the unparalleled generation of human genetic data that needs to be shared on a global level to empower new knowledge, new diagnostics, and new therapeutics for patients and populations. This is at the heart of personalized medicine, according to the alliance.

According to Knoppers, the main challenge is that data from millions of samples may be needed to achieve results and progress showing patterns that would otherwise remain obscure. "That will take new methods and organizational models. Right now, data is typically in silos by type, by disease, by country, by institution; analysis methods

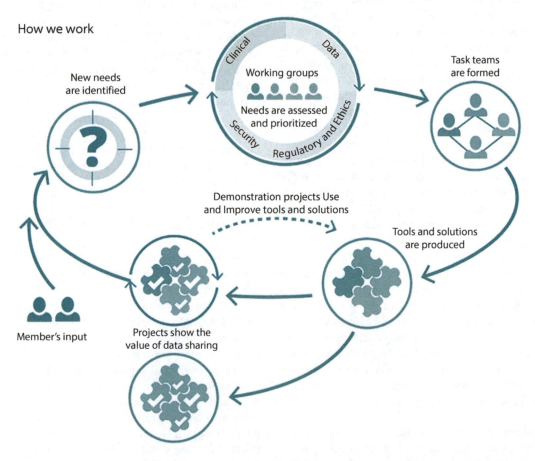

Figure 4.5 How the Genomic Alliance works. Through members input, new needs are identified that are filtered to the working groups that produce new tools and solutions on data sharing. (Courtesy of Bartha Knoppers, 2015. *Personalized and Precision Medicine Conference*, Arrowhead Publishers.)

are nonstandardized, few at scale, and approaches to regulation consent and data sharing limiting interoperability. If we don't act, we risk an overwhelming mass of fragmented data, such as EMRs [electronic medical records] in many countries." Enter the Global Alliance for Genomics and Health, whose members include 416 organizations from 41 countries: universities and research institutes, academic medical centers, life science companies, disease advocacy organizations, and patient groups and professional societies, such as Affymetrix and Roche Diagnostics Canada (Global Alliance for Genomics and Health website). The Global Alliance has four working groups: the Clinical Working Group, to share clinical data and link it to genomic data; the Data Working Group, for data representation storage and analysis of genomic data; the Regulatory and Ethics Working Group, which deals with the ethical, social, and legal implications of the alliance; and the Security Working Group, for data security and privacy protection (Figure 4.5).

While the alliance does not generate data, it has a framework for responsible sharing of genomic and health-related data and a consent policy, with

ways to manage privacy and security risks: the Framework for Responsible Sharing of Genomic and Health-Related Data is a document developed by the Regulatory and Ethics Working Group of the Global Alliance that sets forth a harmonized and human rights approach to responsible data sharing through "foundational principles" and "core elements" (Global Alliance for Genomics and Health website). Additionally, the Public Population Project in Genomics and Society (P3G) is an international not-for-profit consortium dedicated to the development of a multidisciplinary policy infrastructures and research consortium. Building on its international expertise, P3G launched the International Policy Interoperability and Data Access Clearinghouse (IPAC) in 2013. As shown in Figure 4.6, IPAC is a one-stop service for researchers that allows for the development of project-specific ethics-related policies and procedures; review of data and samples access requests, to authorize access to controlled-access databases; and the DataTrust, to support the process of returning individual-level results.

The initiatives by the National Academy of Sciences and Genomic Alliance point to

Figure 4.6 IPAC: The framework for the sharing and release of data. (Courtesy of Bartha Knoppers, 2015. *Personalized and Precision Medicine Conference*, Arrowhead Publishers.)

international ventures to advance precision medicine through diverse mechanisms. The National Academy of Sciences would like to reclassify diseases and create a knowledge network of these reclassifications according to omics categories for prevention and cures. The Genomic Alliance performs data-sharing tools, with working groups hoping to solidify knowledge bases. Each of these initiatives is based on alliances that work on knowledge infrastructures that will catalyze advances in precision medicine. It remains in future hands how these initiatives will turn out and their potential for personalized medicine.

REFERENCES

Chen, R., and M. Snyder. 2013. Promise of personalized omics to precision medicine. *Wiley Interdiscip Rev Syst Biol Med* 5:73–82.

Chowdhury, S. et al. 2015. Do health professionals need additional competencies for stratified cancer prevention based on genetic risk profiling? *Journal of Personalized Medicine* 5:191–212.

Committee on a Framework for Developing a New Taxonomy of Disease, National Research Council. 2011. *Toward Precision Medicine: Building a Knowledge Network for Biomedical Research and a New Taxonomy of Disease.* Washington, DC: National Academies Press.

Dunn, W.B. et al. 2007. Serum metabolomics reveals many novel metabolic markers of heart failure, including pseudouridine and 2-oxoglutarate. *Metabolomics* 3:413–426.

Global Alliance for Genomics and Health website. http://genomicsandhealth.org.

Kim, D., Masyn, K.E., Kawachi, I., Laden, F., and G.A. Colditz. 2010. Neighborhood socioeconomic status and behavioral pathways to risks of colon and rectal cancer in women. *Cancer* 116:4187–4196.

Mirnezami, R., Nicholson, J., and A. Darzi. 2012. Preparing for precision medicine. *New England Journal of Medicine* 366:489–491.

Quinn, K., Kaufman, J.S., Siddiqi, A., and K.B. Yeatts. 2010. Stress and the city: Housing stressors are associated with respiratory health among low socioeconomic status Chicago children. *Journal of Urban Health* 87:688–702.

Sudmant, P.H. et al. 2015. Global diversity, population stratification, and selection of human copy-number variation. *Science* 349:6253.

Great strides in precision medicine: Personalized oncology and molecular diagnostics

In 2011, Matt was a medical student diagnosed with advanced stage IV lung cancer just before approval of Xalkori, a targeted therapy for lung cancer patients harboring the anaplastic lymphoma kinase (ALK) mutation. Matt came from a family of Internet surfers, and his family agreed that it would not be worthwhile to take the additional time to test for the ALK mutation, and Matt was started on Xalkori immediately. Matt had a complete response, but then he relapsed and underwent chemotherapy. The oncologists reintroduced Xalkori going back and forth with precision medicine, but the tumor rapidly progressed. Pfizer, the makers of Xalkori, raised a second-generation drug, which was just starting phase I studies. Matt came into the study and had a complete response. Now, Matt has gone into medical research to do good for others with cancer disease.

This is just one instance of the great successes of precision medicine in oncology, a field where personalized medicine and molecular diagnostics have made extraordinary gains, leading to higher overall survival rates, lower rates of relapse and remission, and more effective treatments with less side effects and adverse drug reactions. Targeted therapy in oncology is based on driver mutations: identification of driver mutations for each cancer is crucial for the development of personalized therapy and molecular tests for diagnosis, monitoring, and management of cancer. The challenges for targeted therapy in cancer are that cancers are heterogenous and difficult to biopsy, and old histological classifications have become outdated in the face of molecular data.

This chapter details the major breakthroughs in personalized oncology in the most prevalent cancers in the United States: lung cancer, breast cancer, colon cancer, melanoma, prostate cancer, brain cancers, and blood cancers. Specifically, the biomarkers (mainly genetic) used to screen for each cancer and the discovered somatic mutations used to determine targeted treatment are underscored here. The targeted therapies emerging from genomics, combined with immunotherapy, have contributed to a molecular revolution in the treatment of cancer, leading to higher survival and lower remission rates. Gleevec is the first-generation targeted treatment for chronic myeloid leukemia (CML), and it was outlined in Chapter 1. Gleevec targets the BCR-ABL mutation, which produces the abnormal tyrosine kinase protein in cancerous blood cells. Its rapid clinical trials, as well as drug resistance in cancer patients, are characteristic of the personalized oncology drugs outlined here. Figure 5.1 gives indications of the timeline for the development of cancer drugs, including BCR-ABL inhibitors, Herceptin, and BRAF mutant inhibitors.

According to Egalite et al. (2014, 660), in a summary on personalized medicine in oncology:

Several approaches have been used over the years in the quest for treating

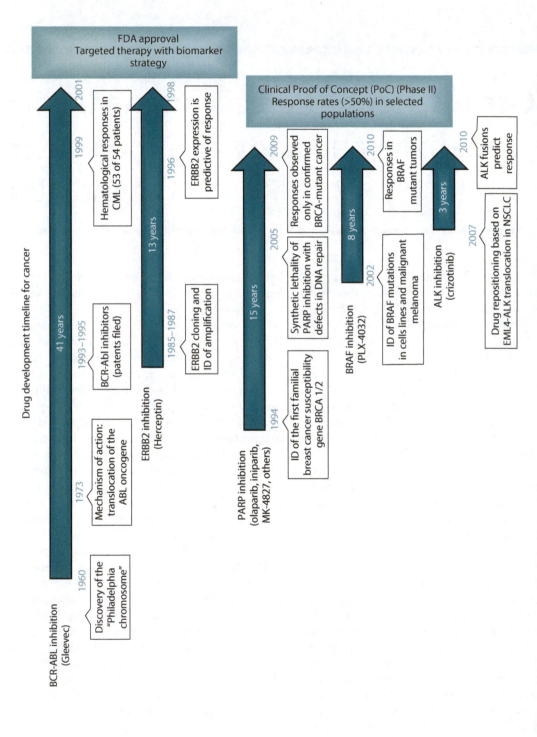

Figure 5.1 Timeline for the drug development of targeted therapies for mutations. PoC. (Reprinted with permission from Chin, L., Andersen, J.N., and P.A. Futreal. 2011. *Nature Medicine* 17:297–303.)

cancer, including surgery, radiation and an array of pharmacological treatments. The application of chemotherapy protocols has been widely used, showing, in many cases, equally poor treatment responses and patient quality of life, with the latter being due to the medications' secondary effects. There are uneven results when applying similar therapies for different and related types of tumors, showing that cancer treatments seem to be particularly suited to a more personalized approach.

In short, previous treatments for cancers were deemed inadequate due to side effects and response, and personalized approaches became necessary.

There is a history of stratified and even personalized management of cancer found in the ways that cancers have been classified based on type, stage and tumor subtype. Current personalized management denotes a variety of methods comprising the identification of cancer risk in a population, testing of biomarkers in order to identify proper therapy (targeted therapy), recognizing the genes linked to drug response (pharmacogenetics/genomics) and analyzing tissues in order to detect recurrence before developing physical symptomatology. The molecular characterization of a tumor in order to select the most appropriate therapy together with a pharmacogenomic analysis denotes that thinking of the patient from a genome-based perspective has long been taking place (Egalite et al., 2014, 660).

We have been contemplating genomic and molecular analysis of cancer tumors based on somatic mutations for a while. This type of genomic analysis is a process that involves identifying the cancer risk, biomarker discovery, pharmacogenomics testing, and tissue analysis.

The evolution of personalized cancer treatment could be exemplified by the management of newly diagnosed patients with chronic myeloid leukemia. The treatment has moved from the general approach using hydroxyurea, IFN-α or allogeneic stem cell transplantation to highly targeted therapies with tyrosine kinase inhibitors (TKIs). Interestingly, while imatinib is the first TKI therapy for chronic myeloid leukemia, approximately 20% of patients do not respond to this treatment, probably due to more "personal" (individual) mutations in the BCR-ABL oncogene, which are currently driving the evolution of second-line TKI therapy (Egalite et al., 2014, 660).

While cancer treatment has moved from traditional therapies, drug resistance has marked first-line agents of targeted therapies due to mutations in oncogenes and tumor suppressors (see below), driving the evolution of second-line therapies.

Meyer et al. defined personalized medicine as a "comprehensive, prospective approach to prevent, diagnose, and treat disease by using each person's unique clinical, genetic, genomic, and environmental information." Some examples of current pharmacological personalized medicine therapies in oncology clinical practice in North America are the use of trastuzumab (breast cancer), imatinib (chronic myeloid leukemia), panitumumab (colorectal cancer), vemurafenib (malignant melanoma) and crizotinib (large-cell lymphoma and non-small-cell lung cancer). These and similar therapies based on determining the patient's molecular and genetic cancer markers help to provide refined treatment decisions for those affected with life-threatening conditions. The characterization of patients' molecular and genetic disease profiles is achieved with "companion diagnostic tests," which are categorized as prognostic, predictive or both. These tests comprise technologies that identify changes in DNA and RNA, epigenetic modifications, altered signaling pathways and protein and metabolic tumor biomarkers. Their use requires evaluation, quality control, standardization and approval from health regulatory organizations. Assessments leading to treatment selection that have been of predictive potential for specific patients when accurate and standardized "allow for proper classification of patients" and for disease management geared towards personalization (Egalite et al., 2014, 660).

Many targeted agents are now in place for breast cancer, CML, colorectal cancer (CRC), melanoma, and lung cancer that are based on biomarkers. These biomarkers are determined by companion diagnostic tests, technologies that isolate DNA and RNA changes, chemical alterations to DNA, and changes in tumor biomarkers.

According to Wijesinghe and Bollig-Fischer (2016, 1), who write in *Lung Cancer and Personalized Medicine*:

Carcinogenesis and the course of the disease for each patient are influenced by many factors including ancestral genetics or germ-line polymorphisms and behavioral or life-style issues. But ultimately cancer is a disease dictated by somatic mutations. Decades of research has contributed to the understanding that cancer initiation and progression are governed by the activation of cancer driver genes, termed oncogenes, and inactivation of key tumor suppressor genes. The importance of oncogenes is underscored by the progress made in developing molecularly targeted drugs to block the function of oncogenes, often proteins with kinase function such as the epidermal growth factor receptors EGFR and HER2. There is a fundamental distinction for activating mutations arising in oncogenes compared to other mutations that are termed passenger mutations. If the mutations confer a selective growth advantage to the cancer cells they are considered to be driver oncogene mutations. Molecularly targeted therapy exploits tumor dependence on activation of driver oncogenes.

Although tumor suppressors are not as directly amenable to targeted therapy, other therapeutic avenues are being explored.

Oncogenesis, as described below, is the key driver for the activation of cancer genes against which targeted therapies are directed, even though lifestyle may be an influencing factor in cancer. While activating oncogenes are aggressively targeted, inactivated tumor suppressors are not as responsive.

Madhu Kalia of Jefferson Medical College, while describing the benefit of introducing targeted therapeutics into clinical oncology practice, points to the clinical utility of using targeted therapies for being able to prospectively characterize patients who likely react to treatment:

The introduction of targeted therapeutics into clinical oncology practice has created major opportunities for further development of ... biomarkers, a process which could evolve into "companion diagnostics." The approvals of trastuzumab, for the treatment of "HER2 overexpressing" atypical breast cancer and imatinib, for the treatment of chronic myelogenous leukemia featuring a *bcr/abl* translocation, and gastrointestinal stromal tumors with selective *c-KIT* oncogene activating mutations, led directly to the development of anticancer therapy focused on the genetic targets on the cancer cells. However, due to biological heterogeneity, only a small number of patients with a particular type of malignancy actually benefit from a specific treatment. In fact, response rates for patients with different types of advanced cancer to currently available drugs varies from about 10% to >90%. Many of the newer biological (targeted therapies), have efficacy in only a minority of patients. This demonstrates the clinical value of being able to prospectively identify patients likely to respond to a specific treatment (Kalia, 2013, S13).

Acquired genetic mutations underlie the pathogenesis of most human cancer. Gene and chromosome abnormalities observed in cancer include gene mutations, chromosome structural abnormalities (translocations, deletions, and insertions), and chromosome number abnormalities (aneuploidy and polysomy). To date, at least 138 known cancer-related genes have been discovered: 74 tumor suppressors and 64 oncogenes that drive tumor growth through 12 known cellular signaling pathways.

CANCER CARE

This new paradigm in oncology care will require an institutional healthcare team comprised of the oncologist, surgeon, primary care physician, and above all, diagnostic pathologist.

> Targeted drugs, in particular therapeutic antibodies, kinase inhibitors and PARP inhibitors, have started to transform current treatment strategies in oncology and new organizational structures.... Prior to application almost all of them require a so called pre-therapeutic companion diagnostic test to identify certain molecular alterations. More precisely, a well-defined biomarker, most often a characteristic genetic alteration or particular protein (over-)expression, that indicates the efficacy of the respective drug has to be identified in the tumor tissue of individual patients. The tissue-based analysis is mostly done by predictive molecular pathology applying conventional or high-throughput techniques on FFPE tissue. This is the basis of personalized, individualized or precision medicine (Dietel et al., 2015, 417).

Biomarkers hold promise in oncology for selecting treatment and predicting clinical outcomes for interventions. There is a distinction between predictive and prognostic biomarker, as described by Cho et al. (2012, 266):

> A prognostic biomarker is related with a patient's clinical outcome and can be used to select patients for an adjuvant systemic treatment irrespective of the patient response to treatment, whereas a predictive biomarker is related to the patient's response to a particular intervention. According to a U.S. NIH Consensus Conference, "a clinically useful prognostic biomarker must be a proven independent, significant factor that is easy to determine and interpret and that has therapeutic consequences." A prognostic biomarker provides information about the patients' overall cancer outcome irrespective of the therapeutic response. Therefore, a prognostic biomarker can be exploited to select patients for an adjuvant systemic treatment but does not forecast the treatment response.

ONCOGENESIS

Multiple genetic alterations may be necessary for transformation of a cell from a normal to a cancerous state (oncogenesis). Oncogenes are the genes that promote transformation and are often derived from normal proto-oncogenes. Certain viral genes can become oncogenic too. Most proto-oncogenes are involved in cellular growth signaling pathways such as *RAS* and *ErbB2* (*Her2*). Mutation of RAS is very common in pancreatic cancer (90%), and overexpression of Her2 is characteristic of a breast cancer subtype. This category of genes suppresses oncogenesis; therefore, loss-of-function mutations on both alleles are needed for inactivation of the gene. Most tumor suppressor genes are involved in the regulation of cell cycle progression or DNA repair after damage or during replication. Examples of tumor suppressor genes are p53, APC, BRCA1/2, and RB1. Indications of a tumor suppressor gene identification are in both the diagnosis and prognosis of cancer, as well as predicting the susceptibility of the carrier to cancer when the mutation is in the germline; for example, the BRCA1/2 mutation in hereditary breast and ovarian cancer.

Molecular abnormalities in solid tumors

The *HER2/neu* gene encodes a protein from the family of human epidermal growth factor receptors. This gene is frequently amplified in breast cancer cells, resulting in increased amounts of Her2 cell surface protein. *HER2* expressing tumors are sensitive to Herceptin monoclonal antibody therapy. Her2 protein is detected by immunohistochemistry (IHC). *HER2/neu* gene amplification is detected by fluorescence *in situ* hybridization (FISH). The epidermal growth factor receptor (*EGFR*) oncogene encodes another member of the same family of epidermal growth factor receptors. This gene is mutated or amplified in several types of cancer cells. Tumors with activating mutations in *EGFR* are sensitive to TKIs. EGFR protein is detected by IHC. *EGFR* gene and chromosome abnormalities are detected by FISH. *EGFR* gene mutations are detected by polymerase chain reaction (PCR), or direct sequencing.

The Kirsten rat sarcoma viral oncogene (*K-ras*) encodes a central component of cell signaling. Mutations in K-ras are the most common oncogene mutations in cancer. K-ras mutations are associated with tumor malignancy and may affect response to some therapies. K-ras gene mutations are detected by direct sequencing.

The 53 kDa tumor suppressor gene (*TP53*) encodes a transcription factor. *TP53* is mutated in half of all types of cancer. Loss of *TP53* function is an indicator of poor prognosis in colon, lung, breast, and other cancers. Mutant p53 protein is detected by IHC. *TP53* gene mutations are detected by a variety of methods, including direct sequencing.

Genome-based medicine for cancer include EGFR kinase inhibitors to treat cancers with EGFR gene mutations, ALK inhibitors to treat cancers with ALK gene translocations, and specific inhibitors of mutant BRAF to treat cancers with BRAF mutations. It is imperative to determine the presence of the mutation that the drug targets before administrating these drugs. Table 5.1 gives a list of biomarkers and type of malignancy they affect.

Kalia has an excellent synopsis of how molecular understanding of cancer is central to the uniqueness of each person's malignancy and the goals of personalized oncology for individuals who have been identified with genetic variants. Our purpose in precision cancer medicine is to select the targeted therapy at the right dosage; predict the patients who will respond to the targeted therapy with less adverse drug reactions; select less expensive, shorter clinical trials; and reduce the overall price of drug development, thus increasing value.

Personalized oncology includes the concept that each individual solid tumor and hematologic malignancy in each person is unique in cause, rate of progression and responsiveness to surgery, chemotherapy and radiation therapy. Genomic and proteomic technologies have made it possible to sub-classify diseases individually using the knowledge of the molecular basis

Table 5.1 Biomarkers, malignancy, type of therapy, and role of biomarker

Predictive biomarker (receptor)	Malignancy	Type of therapy	Biological role of biomarker
ER	Breast cancer	Hormone	Primary target
HER2 (ErbB)	Breast cancer	Trastuzumab	Primary target
Mutant K-RAS	NSCLC	Gefitinib, erlotinib	Downstream of primary target
Mutant K-RAS, BRAF, PIK3, PTEN	Colorectal cancer	Cetuximab, panitumumab	Downstream of primary target
MGMT	Glioma	Alkylating agents	DNA repair
ERCC1	NSCLC	Platinum agents	DNA repair
CYD2D6*	Breast cancer	Tamoxifen	Drug metabolism
TPMT	ALL	6-Mercapto urine, 6-thioguanine	Drug metabolism
UGT1A1*	Colorectal cancer	Irinotecan	Drug metabolism

Source: Modified from Kalia, M. 2013. *Metabolism Clinical and Experimental* 62:S11–S14.
Note: Biological roles of oncological therapy predictive and putative predictive markers. MGMT, methylguanine methyltransferase; TPMT, thiopurine methyltransferase; ERCC, excision cross-complementing; ALL, acute lymphoblastic leukemia; UGT, uridine glucuronyl transferase.

of cancer. Such knowledge has identified differences in gene sequence and/or expression patterns in a number of solid tumors such as breast cancer (HER2), colorectal cancer (KRAS and BRAF), lung cancer (EGF receptor gene [EGFR]) and melanoma (BRAF), as well as for malignant lymphoma and both lymphoid and non-lymphoid leukemias. The ultimate goal of personalized oncology is [the use of] molecular understanding of disease in order to optimize therapy in patients and direct preventive resources and therapeutic agents at "at-risk" normal individuals who have been identified by genetic sequence variants. Cancer drug development and personalized oncology goals are the following: Select optimal drug targets; select optimal drug dosage; predict which individuals will respond to specific drugs at high rates and who will be less likely to suffer toxic effects; select and monitor patients for shorter less expensive advanced clinical trials; reduce the overall cost of drug development; increase value and provide value to patients (Kalia, 2013, S13).

However, some of the complexities include selecting the appropriate drug dosage with pharmacogenomics testing and finding the right clinical trials for patients.

Rodriguez-Antona and Taron (2015, 207) note some of the concerns Kalia may be alluding to:

A major limitation in pharmacogenomics is the difficulty of accurately establishing the utility of the identified markers/strategies for patients and healthcare systems. The level of evidence required to establish that a marker is clinically useful and should be introduced for routine use has been discussed, and general guidelines have been proposed with special emphasis on omics-based markers. However, the possibility of conducting prospective pharmacogenetic-guided clinical trials is limited, making the approach impracticable for many drugs.

TUMOR SUBTYPES

Breast cancer

INCIDENCE AND MORTALITY

There were 231,840 estimated new cases of breast cancer in 2015, representing 14% of all new cancer cases. A total of 40,290 patients died in 2015, representing 6.8% of all cancer deaths. Between 2005 and 2011, 89.4% survived five years. The number of new cases of female breast cancer was 124.8 per 100,000 per year. The number of deaths was 21.9 per 100,000 per year. Approximately 12.3% of women will be diagnosed with female breast cancer at some point during their lifetime, based on 2010–2012 data. In 2012, there were an estimated 2,975,314 women living with breast cancer in the United States (National Cancer Institute website).

Due to the heterogenic nature of breast cancer, it is important to continually develop diagnostic methods and tools that strive to provide consistent and reliable results for optimal patient-specific care. Different phenotypes, often associated with different biological and clinical behaviors, lead to early diagnostic schemes based on appearance. Newer findings now suggest breast cancer arises from mammary stem or progenitor cells, which then differentiate into various lineages by molecular mechanisms not yet fully understood.

INHERITED BREAST CANCER

BRCA1 and BRCA2 are tumor suppressor genes, and mutations in these genes significantly increase risk of breast cancer. Familial breast cancer accounts for 5%–10% of all cases, and BRCA1/2 mutation is present in 20% of familial cases. The average cumulative risk of breast cancer for mutation in either gene is about 27% to age 50 and 64% to age 70. Environmental and genetic factors play a role in the development of cancer in individuals with BRCA mutations. BRCA1/2 are tumor suppressor genes requiring inactivation of both alleles for progression to cancer.

The BRCA1 gene is located on chromosome 17 with 24 exons, while BRCA2 is on chromosome 13 and has 26 exons. Mutations in BRCA genes are heterogeneous, and more than 1000 mutations have been observed. The majority of disease-associated mutations of BRCA1/2 cause protein truncation. DNA sequencing is used for precise

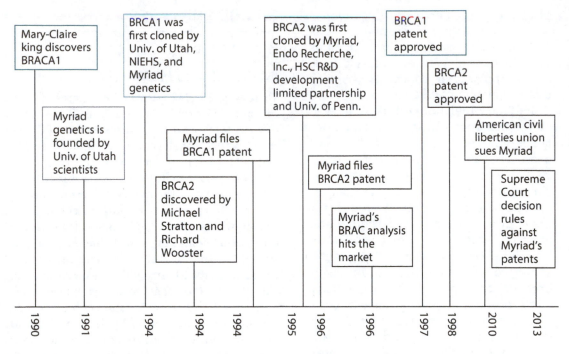

Figure 5.2 BRCA gene life cycle. (Courtesy of http://hubpages.com/hub/Myriad-BRCA12-patents-A-thin-line-between-commerce-consumer-health.)

sequencing of the mutations. Figure 5.2 shows the BRCA gene life cycle.

CURRENT BREAST CANCER DIAGNOSIS STANDARDS

Early classification of stage was based on lymph node status and tumor size, and grades were based on proliferative activity (mitotic index) and degree of differentiation. All play a central role in determining prognosis and treatment course by the physician. Estrogen receptor (ER) and progesterone receptor (PR) status are the gold standard in prescribing systemic adjuvant therapy. Testing for amplification of HER2/neu has become routine for prescribing trastuzumab.

ER, PR, and HER2 status may seem trivial and often conflicts with classification schemes, but it is important for targeted therapy such as tamoxifen and Herceptin. Molecular diagnosis is currently available through various platforms, including MammaPrint, Veridex, Theros, and Oncotype DX, which have improved the resolution of diagnosis. To understand the basis of these tests, one needs to grasp the concepts of histological grading and ER, PR, and HER2/neu status.

MAMMAPRINT

The original datasets were samples from patients younger than 55 years old with tumor size smaller than 5 cm and lymph node negative. Investigators analyzed the gene expression signature of breast cancer patients who developed metastasis within five years of diagnosis compared with those who were metastasis-free during the same time. The analysis identified a 70-gene expression signature that could provide a predictive value for five-year metastasis. This 70-gene expression signature has been shown to outperform traditional clinical and histological parameters for providing a prognostic value.

VERIDEX ROTTERDAM SIGNATURE 76 GENES

In an original study in 2005, investigators separated 115 lymph node-negative breast cancer samples from systemic therapy-naive patients into 35 ER negative and 80 ER positive for analyses. ER-positive samples with distant metastasis had 16 and ER-negative samples with distant metastasis had 60 genes differentially expressed. So, these 76 genes were considered a signature predicting risk of metastasis. A later study only confirmed the prognostic value of this signature but not the independent predictive value for ER-negative patients.

PROSIGNA, PAM50-BASED SIGNATURE

The Prosigna Breast Cancer Prognostic Gene Signature Assay is a qualitative *in vitro* diagnostic tool that utilizes gene expression data weighted together with clinical variables to generate a risk category and numerical score to assess a patient's risk of distant recurrence of disease at 10 years in postmenopausal women with node-negative (stage I or II) or node-positive (stage II), hormone receptor-positive breast cancer. The Prosigna algorithm uses a 50-gene expression profile to assign breast cancer to one of four PAM50 molecular subtypes determined by the tumor's molecular profile. The Prosigna-based signature was Food and Drug Administration (FDA) approved in 2013.

The Oncotype DX platform utilizes quantitative real-time PCR (RT-qPCR) to analyze the expression of 16 cancer-related genes: Ki67, STK15, survivin, CCNB1, MYBL2, GRB7, HER2, ER, PGR, BCL2, SCUBE2, MMP11, CTSL2, GSTM1, CD68, and BAG1, in addition to five reference genes (ACTB, GAPDH, RPLPO, GUS, and TFRC). Oncotype DX predicts risk of recurrence in ER-positive, lymph node-negative breast cancer patients and provides treatment guidelines via a quantitative system of continuous recurrence score (RS) that ranges between 0 and 100 to predict the risk of recurrence within 10 years. This score has also been shown to be an independent predictor of outcome in multivariate survival analyses. The test utilizes RNA extracted from formalin-fixed paraffin-embedded ER-positive breast cancer samples.

Breast cancer patients who are ER positive with a low RS tend to have a low risk of recurrence and will not benefit much from chemotherapy. On the other hand, breast cancer patients who are ER positive with a high RS tend to have a high risk of recurrence and can benefit from chemotherapy. This justifies the use of endocrine therapy in addition to chemotherapy for ER-positive patients with a high RS score.

CONCLUSIONS

Diagnosis of breast cancer has evolved from observing the physical appearance of the tumor under the microscope to quantifying the expression of genes at the molecular level. Although the latter method may seem less subjective and more quantitative, it is still subject to variations, including but not limited to sample handling and reagent lot variations, among other factors. Molecular diagnosis should probably be used in a combinatorial manner, although the cost of doing so might not justify its implementation. Furthermore, next-generation sequencing technologies will likely drive further changes in molecular diagnosis of not only breast cancers but also other diseases.

The Mayo Clinic is running a trial called Breast Cancer Genome Guided Therapy (BEAUTY) that integrates genomics into breast cancer patients. Surgery alone will cure patients since there is a subset that develops metastasis. However, the molecular-derived subtypes of cancer, luminal A, luminal B, basal triple-negative, and HER2+ breast cancers have recently arisen in the classification scheme. As of 2014, we have had better drugs, and within five years, 80% of 5000 patients with an adriamycin, cyclophosphamide, and docetaxel cocktail were still alive. BEAUTY uses patient-derived xenographs where biopsies are injected into mice and tumor behavior is monitored for resistance. Driver targets associated with malignant phenotype are identified as biomarkers.

Colorectal cancer

INCIDENCE AND MORTALITY

There were an estimated 132,700 new cases of CRC in 2015, which represented 8% of all new cancer cases. A total of 49,700 patients died of CRC in 2015, representing 8.4% of all cancer deaths; 64.9% survive five years. The number of new cases per 100,000 is 42.4 for women and men per year. The number of deaths was 15.5 per 100,000 men and women per year. These rates are age adjusted and based on 2008–2012 cases and deaths. Approximately 4.5% of men and women will be diagnosed with colon and rectal cancer at some point during their lifetime. In 2012, there were an estimated 1,168,929 people living with CRC in the United States (National Cancer Institute website).

CRC is the third most common cause of cancer mortality in women and fourth in men. It is fitting therefore that The Cancer Genome Atlas (TCGA) published a comprehensive characterization of the genetics of CRC as its third publication. Because CRC is often diagnosed at a late stage, and the early detection of cancer dramatically increases survival, the identification of genetic risk factors in CRC is of utmost importance. The most widely recognized pathway describing CRC progression is the adenoma carcinoma sequence: beginning as benign polyps or dysplastic lesions before progressing to advanced adenoma, and finally to invasive carcinoma. A variety of molecular pathways and genes are involved in the progression, which typically occurs over years or decades. Carcinomas that remain confined to the colon wall are curable with surgery, while most (73%) of those that progress to stage III tumors (metastasize to regional lymph nodes) are treatable with a combination of surgery, chemotherapy, and radiotherapy.

MOLECULAR PATHOLOGY AND DIAGNOSTICS OF COLORECTAL CANCER

Until recently, the establishment of a CRC diagnosis and prognostic indicator analysis depended on hemotoxylin and eosin staining supplemented with immunohistochemical analysis. Within the past few years, molecular genetic testing has become increasingly important in the diagnosis and management of hereditary CRC syndromes and in determining the treatment options for *de novo* metastatic CRC (mCRC). Five percent of CRCs in the United States are associated with inherited mutations.

The molecular mechanism of sporadic and hereditary CRC is involved in two distinct pathways: (1) chromosomal instability (CIN) and (2) microsatellite instability (MIS). The CIN pathway starts with the loss of function of the APC tumor suppressor gene, mostly with somatic mutation in one allele, followed by chromosomal deletion in a second allele. The MSI pathway arises from mutation in genes involved in DNA mismatch repair (MMR). As a result, DNA replication errors accumulate, especially within microsatellite repeats.

INHERITED CRC SYNDROME

Inherited CRCs arise from mutation in one of the genes involved in the CIN or MSI pathways. Two common inherited CRCs are familial adenomatous polyposis (FAP) and hereditary nonpolyposis colorectal cancer (HNPCC). FAP accounts for less than 1% of CRCs. It is characterized by hundreds to thousands of polyps in the large bowel. Twenty-five percent of cases have no family history, which indicates the presence of new mutations.

HEREDITARY NONPOLYPOSIS COLORECTAL CARCINOMA (LYNCH SYNDROME)

HNPCC is the most common form of hereditary colon cancer and accounts for about 2%–3% of CRCs. HNPCC is characterized by a few polyps that have a potential rapid transformation to carcinoma (one to two years). HNPCC individuals have a lifetime risk of 70%–80% of developing CRC. Ninety percent of HNPCC mutations are observed in genes *MLH1* and *MSH2*. The mutations are located throughout these genes and are diverse. Almost all errors made during replication are proofread by DNA polymerase; the remaining uncorrected mismatched bases will be repaired by the MMR system before cell division. Microsatellites (short tandem repeats) are sensitive to errors during DNA replication. Some targeted genes with repetitive sequences in their coding region that are affected in individuals with MSI include TGFβ receptor, IGF receptor, BAX, MSH6, and MSH3. Eighty-five to ninety percent of HNPCC tumors have MSI. MSI is the production of new alleles from unrepaired replication errors, as shown in Figure 5.3.

Testing of tumor tissue for MSI is the initial lab test to identify HNPCC because MSI is a measure of MMR deficiency. MSI is analyzed by assessing the stability of at least five microsatellite markers, as recommended by the National Cancer Institute.

DE NOVO COLORECTAL CANCER

EGFR is overexpressed in a variety of tumors, including ~70% of individuals with mCRC, with its overexpression linked to a poorer prognosis. EGFR activation by ligand binding eventually activates KRAS and subsequently BRAF, PI3K, and other regulators of cell growth, differentiation, and proliferation. Recently, two anti-EGFR monoclonal antibodies, panitumumab (Vectibix) and cetuximab (Erbitux), have demonstrated efficacy in the treatment of mCRC. In colon cancer, the KRAS mutation determines the response to EGFR therapy, as shown in Figure 5.4.

Activating *KRAS* mutations are strongly linked to the loss of the anti-EGFR antibody response. There are currently three common methods for the detection of *KRAS* mutations:

1. Amplification-resistant mutation system (ARMS) for identifying specific mutations
2. Various sequencing methods
3. High-resolution melting (HRM) analysis

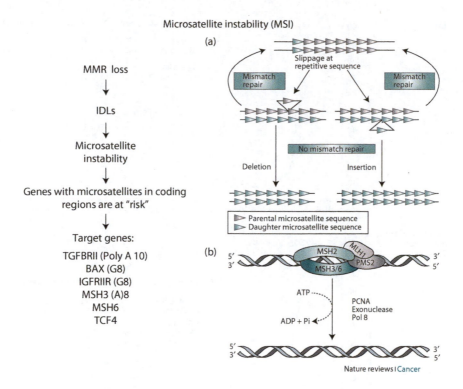

Microsatellite instability (MSI)

(a)

MMR loss
↓
IDLs
↓
Microsatellite
instability
↓
Genes with microsatellites in coding
regions are at "risk"
↓
Target genes:

TGFBRII (Poly A 10)
BAX (G8)
IGFRIIR (G8)
MSH3 (A)8
MSH6
TCF4

Slippage at
repetitive sequence

Mismatch
repair

Mismatch
repair

No mismatch repair

Deletion   Insertion

▷ Parental microsatellite sequence
▷ Daughter microsatellite sequence

(b)

MSH2
MLH1
PMS2
MSH3/6

ATP

PCNA
Exonuclease
Pol δ

ADP + Pi

Nature reviews I Cancer

Figure 5.3 MSI is the production of new alleles from unrepaired replication errors and target genes. IDLs; insertion/deletion loops. (From Colon Cancer Can be Prevented. *Cancer Hub* 3/2014 [http://slideplayer.com/slide/6590729/].)

Each of these techniques has specific advantages and disadvantages, and they are often combined. Direct DNA sequencing is commonly used in KRAS mutation analysis for mCRC. DNA is purified and PCR amplified spanning exons 2 and 3 of the KRAS. The amplicon is purified and sequenced. Sanger sequencing is considered the gold standard and requires at least 20% mutant KRAS DNA for detection in the total sample. Most automated sequencers employ software programs to assist in the analysis. The cobas KRAS Mutation Test (Roche), for use with the cobas 4800 System (Roche), is an allele-specific Taqman probe qPCR test intended for the identification of mutations in codons 12, 13, and 61 of the KRAS gene in CRC.

Melanoma

INCIDENCE AND MORTALITY

There were 73,870 estimated new cases of melanoma in 2015, which represented 4.5% of all new cancer cases. A total of 9940 patients died of melanoma in 2015, which represented 1.7% of all cancer deaths; 91.5% survive five years. The number of new cases of melanoma of the skin was 21.6 per 100,000 men and women per year. The number of deaths was 2.7 per 100,000 men and women per year. Approximately 2.1% of men and women will be diagnosed with melanoma of the skin at some point during their lifetime, based on 2010–2012 data. In 2012, there were an estimated 996,587 people living with melanoma of the skin in the United States (National Cancer Institute website).

Melanoma patients with BRAF mutations respond to treatment with vemurafenib, thus creating a need for accurate testing of BRAF status. Formalin-fixed, paraffin-embedded melanoma samples are macrodissected before screening for mutations using Sanger sequencing, single-strand conformation analysis (SSCA), HRM analysis, and competitive allele-specific TaqMan PCR (CAST-PCR). A concordance of 100% was observed between the Sanger sequencing, SSCA, and HRM techniques. CAST-PCR gave rapid and accurate results for the common V600E and V600K mutations; however, additional assays are required to detect rarer BRAF mutation types found in 3%–4% of melanomas. HRM and SSCA, followed

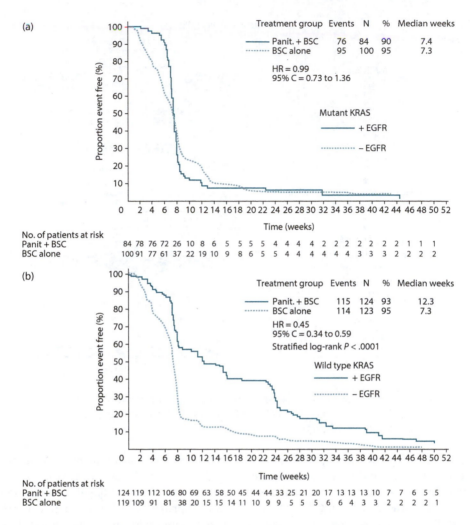

Figure 5.4 KRAS mutation in CRC. In colon cancer, the KRAS mutation determines the response to EGFR therapy. (Reprinted with permission from Amado, R.G. et al. 2008. *Journal of Clinical Oncology* 26:1626–1634.)

by Sanger sequencing, are effective two-step strategies for the detection of BRAF mutations in the clinical setting. A total of 24 V600E, 18 V600K, 4 K601E, 2 V600E2, and 1 V600R mutations were detected.

Circulating DNA from liquid biopsies (LBs) in one study revealed BRAF mutations in melanoma that were validated in patient-derived xenografts, revealing opportunities for alternatives to tissue biopsies, which can be invasive and reveal only one point in time for cancer status. They describe a "powerful technology platform for precision medicine of melanoma patients" using whole exome sequencing and LB of circulating tumor DNA (ctDNA) for patients hard to biopsy. Their results

show hypothesis-driven therapeutic strategies for BRAF mutations and BRAF-resistant melanomas (Girotti et al., 2016).

Lung cancer

INCIDENCE AND MORTALITY

There were an estimated 221,200 new cases of lung cancer in 2015, representing 13.3% of all new cancer cases; 17% of patients survive five years. There were an estimated 158,040 deaths in 2015, representing 26.8% of all cancer deaths. The number of new cases of lung cancer was 58.7 per 100,000 men and women per year. The number of deaths was 47.2 per 100,000 men and women per year.

Approximately 6.6% of men and women will be diagnosed with lung cancer at some point during their lifetime based on 2010–2012 data. In 2012, there were an estimated 408,808 people living with lung cancer in the United States (National Cancer Institute website).

Lung cancers are in the process of being categorized from histological classification to molecular grade based. The majority of lung cancers are non-small-cell lung cancer (NSCLC). Targeted therapies such as gefitinib and erlotinib are in use, and another 25%–30% of patients can enroll in clinical trials targeting other oncogenic drivers. The major lung oncogenic mutations are EGFR, KRAS, and BRAF, and translocations such ALK and ROS1. Tumor suppressor loss-of-function mutations include TP53 and PTEN, which are difficult to target. EGFR is one of the main targets for therapeutic approaches in lung adenocarcinoma. As Wijesinghe and Bollig-Fischer (2016, 2–3) of the Department of Oncology at Wayne State University write:

> By far the majority of lung cancers are categorized as non-small cell lung cancer (NSCLC) and about 15% minority are small cell lung cancer. NSCLCs are further subdivided into adenocarcinomas (~45%), squamous cell lung cancer (~23%), and large cell lung cancer (~3%), with other subtypes representing the remaining approximate 28%. During the last decade there has been a shift in classification of lung cancer based on tumor genetics. This attempt not only provided the actionable targets for the effective therapy but highlighted the importance of reconsidering the tumor reclassification, from histology based to molecular based.... Of the three major subtypes of lung cancers, patients with adenocarcinoma benefit the most from molecular genomic based cancer therapeutics today. While 25%–30% of patients receive targeted therapies like gefitinib and erlotinib, another 25%–30% can enroll in clinical trials targeting other known oncogenic drivers. The major oncogenic drivers in lung adenocarcinoma include activating mutations in EGFR, KRAS, BRAF, HER2,

> MET or translocations of ALK, ROS1 and RET; and all of these targets have drugs that are approved or in clinical trials. Tumor suppressor loss-of-function mutations occurring in lung adenocarcinoma include TP53, CDKN2A, PTEN, STK11, RB1, NF1, KEAP1 and SMARCA4. Targeting these tumor suppressor alterations is therapeutically challenging at the moment but their presence may be highly informative such as in the case of TP53 mutation association with lack of response to EGFR inhibitors and recurrence.

Cagle et al. (2016) describe the two major predictive biomarkers, EGFR mutations and ALK translocation, which have been validated in clinical trials and approved by the FDA. However, therapies directed against these biomarkers have developed resistance, and investigators have not yet discovered targets for additional biomarkers, such as KRAS, the most frequent oncogenic driver.

> Two predictive biomarkers for personalized therapy of non-small cell lung cancers (NSCLC) have been well validated in clinical trials and approved by the Federal Drug Administration (FDA): epidermal growth factor receptor (EGFR) mutations and anaplastic lymphoma kinase (ALK) translocations. These two biomarkers have been the subject of the first lung cancer biomarkers guidelines from the College of American Pathologists (CAP), International Association for the Study of Lung Cancer (IASLC) and Association for Molecular Pathology (AMP) as well as the CAP Lung Cancer Biomarker Reporting Template. The frequency of EGFR mutations found in non-small cell lung cancers (NSCLC), more specifically in adenocarcinomas, ranges from about 15% of whites and Hispanics to about 19% of African Americans to about 30% of Asian patients. ALK translocations creating fusion genes occur in about 4%–5% of adenocarcinomas. Lung cancers that initially respond to

first generation EGFR TKIs or to crizo-tinib eventually develop drug resistance and relapse, typically within a year. Since about 80% of adenocarcinomas lack EGFR mutations or ALK transloca-tions and since lung cancers with these abnormalities develop acquired resis-tance to current therapies, there has been a robust search for additional oncogenic drivers in lung cancers that might be actionable. Investigations have not yet discovered drugs that tar-get KRAS, the most frequent oncogenic driver in lung adenocarcinomas, occur-ring in about 30% of cases. Oncogenic drivers have not yet been identified in a substantial number of lung adenocar-cinomas and, of the additional drivers that have been identified, investigations of several are sufficiently advanced that they are being considered for revisions to the CAP/IASLC/AMP lung cancer bio-marker guidelines and CAP lung cancer biomarker reporting template (Cagle et al., 2016, 26).

As early as 2014, personalized cancer medicine made extraordinary gains for targeted treatment. According to a review by Okimoto and Bivona (2014, 309–310) of the University of California, San Francisco, in *Personalized Medicine* on personal-ized management of lung cancer:

The discovery of genetic alterations that drive tumor progression in subsets of non-small cell lung cancer (NSCLC) has transformed the clinical manage-ment of this disease. In particular, recent therapeutic advances in NSCLC have been established by focusing on unique somatic genetic variations between patients that predict response to targeted therapies. This biomarker-driven paradigm in NSCLC has not only revealed significant interpatient tumor heterogeneity, but also extensive intratumor heterogeneity. While this biological variability poses significant challenges to identify clinically relevant driver genes, our ability to molecularly dissect individual tumor types and classify them based on their genetic profile provides tremendous opportu-nity to rationally design and therapeu-tically target these genetic events to improve patient care. The prototypical example of how genotype directed, biomarker-driven lung cancer care can dramatically change an entire treatment paradigm is exemplified through the discovery of EGFR mutations in NSCLC that predict tumor responsiveness and improve progression-free survival (PFS) during therapy with EGFR tyro-sine kinase inhibitors (TKIs) compared with standard chemotherapy. More recently, the identification of the EML4-ALK gene rearrangement is yet another example of a therapeutic biomarker that has demonstrated tremendous clinical success as a predictive marker of effective ALK inhibitor (crizotinib) treat-ment. These two clinically validated oncogenic drivers exemplify the unique benefits of genotype-directed targeted therapy and demonstrate the impor-tance of patient stratification to enrich and enhance treatment responses in drug development and clinical trials. As the list of potential oncogenic drivers in lung cancer continues to grow, our ability to match an individual patient's tumor profile to a specific targeted therapy hinges upon a comprehensive molecular diagnostic platform that can not only rapidly screen for known tar-getable genes, but also identify novel actionable drug targets that can be therapeutically exploited to improve clinical outcomes.

Phase III clinical trials of the EGFR inhibitor gefitinib identified EGFR mutant-driven lung ade-nocarcimas that led to genotypic stratification of the cancer predominating over clinicopathologic characterization. Clinical trials demonstrated that patients, many of whom are nonsmokers and also EGFR mutant positive, treated with targeted therapy such as EGFR TKIs, had improvement in progression-free survival.

The second driver mutation discovered in NSCLC patients was the EML4-ALK translocation,

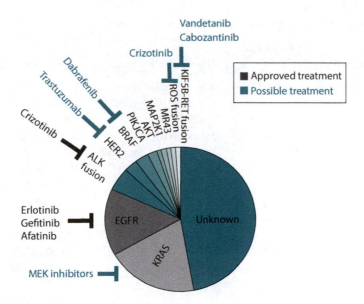

Figure 5.5 Treatments and targeted mutations for lung cancer. (Reprinted with permission from Rodriguez-Antona, C., and M. Taron. 2015. *Journal of Internal Medicine* 277:201–217.)

which encodes for the protein with kinase activity. Preclinical data suggested that ALK rearrangements are oncogenic drivers that are susceptible to targeted therapies in a subset of lung adenocarcinomas. EML4-ALK rearrangements are present in approximately 3%–5% of patients with lung adenocarcinoma. Many of these patients are younger and light smokers. Isolating EML4-ALK rearrangements and EGFR mutations within patient populations is of great importance in genotype-directed clinical trial design, since these mutations are mutually exclusive. According to Okimoto and Bivona (2014, 313), in order to directly evaluate the effect of ALK inhibition in patients with EML4-ALK rearrangements:

A multicenter, single-arm study of crizotinib was conducted in 2010, which has already demonstrated impressive clinical efficacy, with an objective response rate (ORR) of 61% and median duration of response of 47 weeks. These findings served as the foundation for accelerated approval of crizotinib by the U.S. FDA in 2011 for the treatment of advanced ALK-positive NSCLC. More recently, a randomized, multicenter, phase III study of crizotinib versus standard of care chemotherapy in

ALK-positive NSCLC patients who had previously received platinum doublet therapy demonstrated a median PFS of 7.7 versus 3.0 months in the crizotinib.

Xalkori, manufactured by Pfizer, is a drug that targets ALK mutations. Figure 5.5 shows a pie chart of the treatments and targeted mutations for lung cancer, and Table 5.2 demonstrates oncogenic drivers in NSCLC.

TKI therapy should not be administered as first-line therapy for NSCLC patients who have wild-type EGFR, since they respond better to chemotherapy. Single-alteration detection methods approved by the FDA in Clinical Laboratory Improvement Amendments (CLIA)-certified laboratories should be no longer than 10 days (Figure 5.6).

Randomized controlled trials have demonstrated that NSCLC patients with wild-type EGFR respond better to systemic chemotherapy. Therefore, in most clinical circumstances, EGFR TKI therapy should not be administered as first-line therapy without documented evidence of a sensitizing EGFR mutation. Based on these clinical findings, coupled with the fact that patients with stage IV NSCLC have a median survival of approximately

Table 5.2 Targetable oncogenic drivers in non-small-cell lung cancer

Gene	Alteration	Frequency (%)	Targeted therapies	Current clinical trials[a]
EGFR	Mutation	10–15	Erlotinib, gefitinib, afatinib, CO-1686, AZD9291	NCT01836341, NCT01542437, NCT01953913, NCT01931306, NCT01526928, NCT01802S32
BRAF	Mutation	3–4	Dabrafenib, trametinib, dasatinib	NCT01336634, NCT01362296 NCT01514864
PI3KCA	Mutation	1–3	BKM120, XL147	NCT01297452, NCT01570296, NCT01297491, NCT01723800, NCT01390818
HER2	Mutation	1–4	Afatinib, neratinib, dacomitinib	NCT01542437, NCT01827267, NCT01858389, NCT00818441
EML4-ALK, KIF5B-ALK, TFG-ALK	Fusion	3–5	Crizotinib, LDK378, CH5424802	NCT00932451, NCT01639001, NCT01685060, NCT01685138, NCT01828112, NCT01828099, NCT01579994, NCT01871805
ROS1	Fusion	1–2	Crizotinib	NCT01945021
RET	Fusion	1–2	Cabozantinib, vandetanib	NCT01639508, NCT01823068

[a] Trials can be accessed via the ClinicalTrials.gov website.

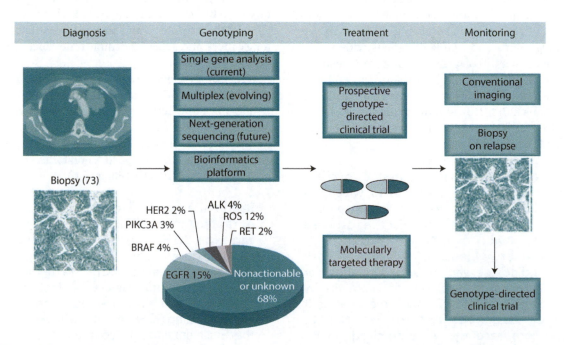

Figure 5.6 Workflow for targeted therapy of lung cancer. (Reprinted with permission from Okimoto, R.A., and T.G. Bivona. 2014. *Personalized Medicine* 11:309–321.)

16 weeks if left untreated, evidence-based guidelines have been established that mandate the turnaround time for EGFR molecular testing to be no longer than 10 days. To meet this need, the majority of Clinical Laboratory Improvement Act-certified laboratories are now implementing FDA-approved mutational detection assays to identify single gene alterations. These detection methods are variable, but are all based on sequencing, amplification or FISH analysis of the mutant EGFR or KRAS allele or ALK rearrangements, respectively (Okimoto and Bivona, 2014, 315).

MOLECULAR DIAGNOSIS

Multiplex genotyping is currently being developed to detect mutational hotspots in multiple targeted genes and validated in clinical trials:

Given the relatively small number of actionable targets in NSCLC, single gene-based molecular tests for EGFR and EML4-ALK continue to be the standard of care. As we continue to clinically validate the growing list of potential driver mutations in lung cancer, however, we will need to develop a more comprehensive, cost-effective tumor genotyping protocol to rapidly screen patients for all actionable therapeutic targets will need to be developed. Multiplex genotyping, a PCR-based assay designed to simultaneously detect the expression of known mutational hotspots in multiple target genes in a single reaction, is currently being employed throughout the cancer research community and being validated in clinical trials (Okimoto and Bivona, 2014, 315).

Personalizing NSCLC treatment through genotype-directed targeted therapy involves

The identification of targetable oncogenic drivers that define molecular subsets in NSCLC [which] has transformed the clinical management of this disease. Initial clinical design based on large, unselected populations failed to show a significant benefit in clinical outcomes in patients with NSCLC treated with targeted therapies. Conversely, when patients are prospectively selected based on their molecular profile and rationally directed to an appropriate targeted therapy, unprecedented results emerge as evidenced by EGFR and EML4-ALK in NSCLC. The remarkable efficacy of these molecularly targeted therapies in selected patient populations has led to a significant surge in the number of genotype-directed clinical trials, heralding the beginning of personalized medicine in lung cancer (Okimoto and Bivona, 2014, 317).

The efficacy of developing a molecular profile for NSCLC patients is tremendously worthwhile, given the value of having directed targeted therapy, especially against EGFR mutations and ALK translocations, even if the design does not significantly advance the clinical outcomes of patients with NSCLC. With the progress of clinical trials and the patient selection, we may have targeted therapies directed against KRAS and clinically validated molecular diagnostics.

Prostate cancer

INCIDENCE AND MORTALITY

There were an estimated 180,890 new cases of prostate cancer in 2016, representing 10.7% of all new cancer cases. There were 26,120 estimated deaths in 2016, representing 4.4% of all cancer deaths. Between 2006–2012, 98.9% of patients survive within five years. The number of new cases of prostate cancer was 129.4 per 100,000 men per year. The number of deaths was 20.7 per 100,000 men per year. Approximately 12.9 percent of men will be diagnosed with prostate cancer at some point during their lifetime based on 2011–2013 data. In 2013, there were an estimated 2,850,139 men living with prostate cancer in the United States (National Cancer Institute website).

A whole genome and exome sequencing study has been conducted in a multisite study published in *Cell* that discovered somatic mutations and structural rearrangements in prostate cancer. The participants were 150 men with metastatic prostatic cancer who developed resistance to primary androgen deprivation therapy. Alterations in TP53 and PTEN were discovered, among other mutations that are clinically actionable. The study investigators recommend that prostate-specific antigen (PSA) be used along with biomarkers to determine clinical course:

> Prostate cancer is among the most common adult malignancies, with an estimated 220,000 American men diagnosed yearly. Some men will develop metastatic prostate cancer and receive primary androgen deprivation therapy (ADT). However, nearly all men with metastatic prostate cancer develop resistance to primary ADT, a state known as metastatic castration-resistant prostate cancer (mCRPC). Multiple "second generation" ADT treatments, like abiraterone acetate and enzalutamide, have emerged for mCRPC affected individuals; however, nearly all affected individuals will also develop resistance to these agents. In the U.S., an estimated 30,000 men die of prostate cancer yearly....
>
> Toward development of a precision medicine framework for metastatic, castration-resistant prostate cancer (mCRPC), [a study led by investigators at the University of Michigan Medical School] established a multi-institutional clinical sequencing infrastructure to conduct prospective whole-exome and transcriptome sequencing of bone or soft tissue tumor biopsies from a cohort of 150 mCRPC affected individuals. Aberrations of AR, ETS genes, TP53, and PTEN were frequent (40%–60% of cases), with TP53 and AR alterations enriched in mCRPC compared to primary prostate cancer. We identified new genomic alterations in PIK3CA/B, R-spondin, BRAF/RAF1, APC, b-catenin, and ZBTB16/PLZF. Moreover, aberrations of BRCA2, BRCA1, and ATM were observed at substantially higher frequencies (19.3% overall) compared to those in primary prostate cancers. 89% of affected individuals harbored a clinically actionable aberration, including 62.7% with aberrations in AR, 65% in other cancer-related genes, and 8% with actionable pathogenic germline alterations. This cohort study provides clinically actionable information that could impact treatment decisions for these affected individuals. However, the currently used clinical prognostic factors of T-category, PSA and Gleason score explain only a moderate proportion of the observed variation in clinical outcome. The use of PSA alone to determine the clinical course in otherwise "clinically silent" disease needs to be buttressed with biomarkers based on tumor biology. GWAS has identified multiple common genetic variants associated with an increased risk of prostate cancer, but these explain less than one third of the heritability. Men who harbor mutations in genes involved in DNA repair and genome surveillance (BRCA2) have an increased risk for prostate cancer and poor prognosis. Whole genome and exome sequencing has identified somatic mutations and androgen driven structural rearrangements (Robinson et al., 2015, 1216).

Brain cancers

INCIDENCE AND MORTALITY

There were an estimated 22,850 new cases of brain cancer in 2015, representing 1.4% of all new cancer cases, with a 33.3% five-year survival rate. The estimated deaths in 2015 from brain cancer were 15,320, representing 2.6% of all cancer cases. The number of new cases of brain cancer was 6.4 per 100,000 men and women per year. The number of deaths was 4.3 per 100,000 men and women per year. Approximately 0.6% of men and women will be diagnosed with brain cancer at some point during their lifetime, based on 2010–2012 data. In 2012, there were an estimated 148,818 people living with brain cancer in the United States (National Cancer Institute website).

According to Ene and Holland (2015, S90) of the University of Washington School of Medicine, there are currently no FDA-approved drugs designed for personalized therapy for patients with gliomas.

Unfortunately, targeting single gene products or alterations may not be feasible in a majority of tumors especially GBMs given significant clonal diversity inherent to this tumor type. This represents one of the most frustrating aspects of research seeking to develop targeted therapy or personalized medicine for GBMs.... Gliomas are classified into four grades by the World Health Organization (WHO) based on pathologic features of such as cellularity, pleomorphism, endothelial proliferation/abnormal angiogenesis, mitotic figures and necrosis. Glioblastoma multiforme (GBM) represents the worst grade of gliomas (Grade IV) and is also the most common form of primary brain tumors in adults. Although the WHO grades have distinct median survival differences between grades (I: 8–10 years, II: 7–8 years, III: 2 years, IV: <1 year), it does not account for the variability in response to therapy within each grade that may be driven by heterogeneity at the molecular and cellular levels. Consequently, the goal of targeted therapy for glioma is to develop a clinically relevant algorithm that predicts response to specific therapy based on patient specific molecular/cellular features.

Table 5.3 shows the frequently mutated genes in glioblastoma.

MOLECULAR CLASSIFICATION OF GLIOBLASTOMA

There are signs, however, that this advance is in the near future. "The DNA alkylating agent Temozolomide (TMZ) improves the survival of patients with GBM when used in combination with radiation therapy. Furthermore, GBMs with hypermethylation and suppression of O-6-methylguanine DNA methyltransferase (MGMT) are more sensitive to the TMZ. MGMT hypermethylation, however, represents a small minority of patients with GBM" (Ene and Holland, 2015, S90). However, according to Ene and Holland:

Given that TMZ plus adjuvant radiation improves survival of patient irrespective

Table 5.3 Molecular classification of glioblastoma multiforme

Subclassification	Frequently mutated genes
Proneural	TP53 (54), IDH1 (30), PIK3R (19), EGFR (16), PDGFRA (11)
Neural	EGFR (26), TP53 (21), PTEN (21), ERBB2 (16), NF-1 (16), PIK3R (11)
Classic	EGFR (32), PTEN (23), EGFRvIII (23)
Mesenchymal	NF-1 (37), TP53 (32), PTEN (32), RB1 (13)

Source: Modified from Ene, C.I., and E.C. Holland. 2015. Surgical Neurology International 6(Suppl. 1): S89–S95.

Note: The numbers in parenthesis are the original references for the discovery of the mutated genes.

of MGMT methylation status and the lack of an alternative agent with clinical efficacy, designing prospective randomized clinical trials where one group of patients receive TMZ and [other alkylating agents] becomes problematic and unethical. So at this point, the sensitivity of MGMT hypermethylated tumors to TMZ only represents proof of concept that supports targeting a sub-set of GBM patients with specific molecular signatures. In the future, as more chemotherapeutic agents with similar efficacy are developed based on molecular alterations, it may be possible to design clinical trials assessing the differential sensitivities of patients with different molecular signatures and alterations to chemotherapy (Ene and Holland, 2015, S90).

TUMOR IMMUNOLOGY

The differential expression of antigens also provides an avenue for personalized medicine for GBMs. Tumor vaccines are being developed from exposure of immune cells to patient tumor antigens, allowing for their administration to selectively target patient tumors, such as stem cells that are hypothetically reconstituted following treatment. Different antigens that express GBM cancer stem cells could be targeted for a patient subpopulation.

Hematologic malignancies

INCIDENCE AND MORTALITY

There were an estimated 54,270 new cases of leukemia in 2015, representing 3.3% of all new cancer cases. A total of 24,450 patients were estimated to have died from leukemia in 2015, representing 4.1% of all cancer deaths; 58.5% survive five years. The number of new cases of leukemia was 13.3 per 100,000 men and women per year. The number of deaths was 7.0 per 100,000 men and women per year. Approximately 1.5% of men and women will be diagnosed with leukemia at some point during their lifetime, based on 2010–2012 data. In 2012, there were an estimated 318,389 people living with leukemia in the United States (National Cancer Institute website).

There were an estimated 71,850 new cases of non-Hodgkin lymphoma in 2015, representing 4.3% of all new cancer cases. A total of 19,970 patients died of non-Hodgkin lymphoma in 2015, representing 3.4% of all cancer deaths; 70% survive five years. The number of new cases of non-Hodgkin lymphoma was 19.7 per 100,000 men and women per year. The number of deaths was 6.2 per 100,000 men and women per year. Approximately 2.1% of men and women will be diagnosed with non-Hodgkin lymphoma at some point during their lifetime, based on 2010–2012 data. In 2012, there were an estimated 549,625 people living with non-Hodgkin lymphoma in the United States (National Cancer Institute website).

Hematologic malignancies are the object of small-molecule inhibitors that target genomic alterations found in cancerous cells, such as copy number variations and point mutations. Companion diagnostic tests that stratify patient populations according to distinctive genetic makeup enable these targeted therapies to intercept specific targets. They, in turn, have fewer side effects and are better tolerated. However, resistance has emerged. According to a review by Gayane Badalian-Very (2014, 73) of the Department of Medical Oncology in the Dana-Farber Cancer Institute:

It is well known that most hematologic malignancies are caused by genomic alteration (point mutation, chromosomal aberrations, copy number variations), and therefore complete understanding of these diseases can only be achieved by comprehensive screening of a large number of clinical samples. Currently there are several small molecular inhibitors in clinical practice. "In the future we should drive our focus on enhancing the patient's response based on their unique genetic makeup using appropriate companion diagnostics along with targeting the driver event". Also, to maximize the benefits of small molecular inhibitors, we must deliver the targeted agents to a susceptible population based on individuals' susceptibility profiles determined by companion diagnostics. Targeted therapies or small molecular inhibitors block the proliferation of cancer cells by intercepting their specific target. Since their range of action is smaller than general chemotherapy agents, the adverse effect caused by these inhibitors is smaller as well and they are better tolerated by patients. This seems to be a success story, but the final picture is complicated. These inhibitors are not curative, and disease relapse remains a fairly common complication in these malignancies. Several reasons could be attributed to disease relapse, among which is the escape of tumors cells which obtain new surviving mutations and the evolution of new neoplastic populations due to weakened immune response. On the other hand, there are obvious caveats in targeting cancer cells with very specific molecular inhibitors.

Table 5.4 shows targeted treatment in hematological malignancies.

Finding mutations for hematologic malignancies is the basis for precision medicine advances in blood cancers. These cancers are usually screened for by PCR-based technology or massively parallel sequencing, such as whole genome sequencing, that sequences the entire genome and can uncover actionable cancer-associated mutations. Drug resistance as a result of compensatory mechanisms through which cancer cells escape inhibition has

Table 5.4 Targeted treatment according to mutation in hematological malignancies

Gene	Genetic alterations	Tumor type	Targeted agent
Receptor tyrosine kinase			
ALK	Mutation, CNV	Anaplastic large cell lymphoma	Crizotinib
FGFR1	Translocation	CML, myelodysplastic disorders	Imatinib methylase
FGFR3	Translocation, mutation	Multiple myeloma (113)	PKC412, BIBF-1120
FLT3	CNV	AML	Lestaurtinib, X1999
PDGFRB	Translocation, mutation	CML	Sunitinib, sorafenib, imatinib, nilotinib
Nonreceptor tyrosine kinase			
ABL	Translocation (BCR-ABL)	CML, AML	Dasatinib, nilotinib, bosutinib
JAK2	Mutation (V6l7F) translocation	CML, myeloproliferative disorders	Lestaurtinib, INCB018424
ERK1/2	Mutation	Mantle cell lymphoma, CLL	Ibrutinib
Serine–threonine kinase			
Aurora A and B kinase	CNV	Leukemia	MK5108
BRAF V600E	Mutation	LCH, ECD (110), hairy cell leukemia (112)	Vemurafenib (PLX4032)
Polo-like kinase	Mutation	Lymphoma	B12536
Nonkinase targets			
PARP	Mutation, CNV	Advanced hematologic malignancies, CLL, mantle cell lymphoma	BMN 673
Antibodies			
CD20		Hodgkin lymphoma	Rituximab
CD52		B-cell chronic lymphocytic leukemia	Alentuzumab
CD20		Non-Hodgkin lymphoma	Ibritumomab tiuxetan
Apoptotic agents			
Proton pump inhibitors		Multiple myeloma, mantle cell lymphoma, peripheral T-cell lymphoma	Bortezomib, pralatrexate

Source: Modified from Badalian-Very, G. 2014. *Computational and Structural Biotechnology Journal* 10:70–77.
Note: CNV, copy number variation; LCH, Langerhans cell histiocytosis; ECD, Erdheim Chester disease.

led to the development of second-generation targeted therapies.

Clinical trials between standard chemotherapeutic agents and targeted therapies such as small-molecule inhibitors vastly differ, according to Badalian-Very (2014, 73):

There are several fundamental differences between cytotoxic chemotherapies and small molecular inhibitors. Dose-related toxicities have traditionally been considered [central] end points of Phase I trials and the maximum

Table 5.5 Anticancer treatments approved by the FDA carrying companion diagnostics

Biomarkers with pharmacokinetic effect	TPMT (mercaptopurine, thioguanine)
	UGR1A1 (irinotecan, nilotinib)
Biomarkers with pharmacodynamic effect	EGFR (cetuximab, erlotinib, gefitinib, panitumumab, afatinib)
	KRAS (cetuximab, panitumumab)
	ABL (imatinib, dasatinib, nilotinib)
	BCR-ABL (bosutinib, busulfan)
	ALK (crizotinib)
	C-kit (imatinib)
	HER2/neu (lapatinib, trastuzumab)
	ER (tamoxifen, anastrozole)

Source: Modified from Badalian-Very, G. 2014. Computational and Structural Biotechnology Journal 10:70–77.
Note: All the genes are used for companion diagnostics of the drugs mentioned in the brackets.

tolerated dose (MTD) was regarded as the optimal dose providing the best efficacy with manageable toxicity. Recently, development of targeted inhibitors has challenged the paradigms used in cytotoxic chemotherapy trial design. In precision medicine pharmacokinetic (PK) and pharmacodynamic (PD) end points tend to take a backseat to toxicity. Molecularly targeted agents do not always maintain the same dose–toxicity relationship as cytotoxic agents and tend to produce minimal organ toxicity. Furthermore, molecular therapeutic agents usually result in prolonged disease stabilization and provide clinical benefit without tumor shrinkage, a characteristic seen with cytotoxic agents, therefore necessitating alternative measures of antitumor efficacy. These end points include biologically relevant drug exposures, PD biomarker measures of target inhibition, and intermediate end-point biomarkers, such as … biomarkers. In the field of cancer, pharmacogenomics is complicated by the fact that two genomes are involved: the germline genome of the patient and the somatic genome of tumor, the latter of which is of primary interest. This genome predicts whether specific targeting agents will have a desired effect in the individual. On the other hand, germline pharmacogenetics can identify patients likely to demonstrate severe toxicities when given cytotoxic treatments (Badalian-Very, 2014, 73).

Cytotoxic dose chemotherapy used to be the paradigm in determining the best efficacy for phase I trials. However, intermediate end points such as biomarkers are now seen as measures of target inhibition since molecular therapies lead to target inhibition. It is the somatic genome of the tumor that is of primary interest. Table 5.5 gives a list of anticancer treatments approved by the FDA carrying companion diagnostics.

TRENDS IN PERSONALIZED ONCOLOGY

MicroRNAs

The microRNA (miRNA) paradigm shifted from worm development to cancer when G.A. Calin et al. identified miR-15 and miR-16 as tumor suppressors in chronic lymphocytic leukemia in 2002. Fifty percent of CLL and MCL patients have homozygous loss of this genetic region (13q14), and 68% of patients do not express miR-15 and 16. Shortly after, Michael et al. (2003) showed a similarly reduced expression of mature miR-143 and miR-145 in CRC. In Takamizawa et al., 2004, showed that let-7 was often lost in lung cancer. These landmark studies established the importance of miRNAs in cancer.

Blenkiron et al. (2007) showed that 133 miR-NAs are differentially expressed in the different molecular subtypes of breast cancer using a bead-based flow cytometric miRNA expression profiling method. These miRNAs include pathologically important miRNAs: members within the let-7, miR-200, and miR-10 family; the miR-17–92 and miR-214 clusters; and individual miRNAs such as miR-155 and miR-21.

Liquid biopsies

Circulating cell-free nucleic acids (ccf-NAs) are derived from dying tumor cells; RNA/DNA from circulating exosomes (exoRNA) is derived from living tumor cells; DNA/RNA from circulating tumor cells (CTCs) are derived from living and functional tumor cells. Circulating DNA, or cell-free DNA, is tumor cells extracted from malignancies. Clinical data has been collected to assess target driver mutations with biomarkers. In one study of 206 patients with metastatic colon cancer, the sensitivity and specificity for ctDNA for the detection of medically relevant KRAS gene mutations were 87.2% and 99.2%, respectively (Bettegowda et al., 2014). According to the same study, in patients with localized tumors, ctDNA was detected in 73%, 57%, 48%, and 50% of patients with CRC, gastroesophageal cancer, pancreatic cancer, and breast adenocarcinoma, respectively.

Circulating DNA constitutes a development of noninvasive methods to detect and monitor tumors. Digital PCR-based technologies exist to evaluate the proficiency of circulating DNA in 640 patients with diverse cancer types. Circulating DNA in tumor cells was found in many malignancies and cancers. Circulating DNA in tumors is part and parcel of broadly applicable biomarkers, the study affirms.

Tissue biopsy has long served as the mainstay of cancer diagnosis, staging, and therapeutic decisions, with its role evolving from simple histologic examination to complex genetic analysis. Despite its utility, biopsy represents only a single time point from a single location, often proving inadequate at fully characterizing a malignancy and its evolution because nearby tissue might contain additional genetic information that would affect staging or treatment. Unfortunately, the invasive nature and inherent selection bias of biopsy limit its usefulness as a real-time monitoring tool. Newer technologies seeking to transcend this shortcoming include analysis of circulating tumor cells and their fragments, such as exosomes and DNA, in peripheral blood. A recent feature in Nature highlights advancements in the detection of circulating tumor DNA (ctDNA) for this purpose. This emerging methodology enables sequencing of DNA originating from lysed tumor cells present in a blood sample to follow tumor recurrence and to characterize genetic abnormalities that confer resistance (Ryder and Schmotzer, 2015, 443).

LBs are needed for several reasons:

- Obtaining tissue biopsies is limited.
- Tissue biopsy is an invasive procedure, and it is not advised for periodic patient monitoring.
- The average cost of a lung cancer tissue biopsy is $14,634.
- In the United States, more than 1.6 million breast biopsies are performed each year.
- LBs are basically blood tests that help in detecting molecular biomarkers in body fluids.

The advantage of LB methods are as follows:

- LB overcomes many of the limitations of tissue biopsy, and it is also an inexpensive procedure.
- LBs may soon become the stethoscope for the next 200 years.
- LB allows us to detect tumors at an early stage and then monitor them for metastasis.
- The tests can also be used to detect whether the tumors would resist cancer treatments.

In considering the validation of LB diagnostic tests, two options have been widely discussed—correlation with results from tissue biopsy and correlation with clinical outcome. Correlation studies with tissue are acknowledged to be imperfect because of differences in biology, as well as test design, but studies correlating with clinical outcome are costly. Regarding validation of LBs, what level of evidence would one require for use? Clinical outcome demonstrating detection of circulating markers indeed correlates with response.

Immunotherapy

Immunotherapy combined with genomic medicine based on cancer mutations is rapidly becoming the standard of care in personalized oncology, for example, pembrolizumab (Keytruda) and ipilimumab (Yervoy) for treatment of advanced melanoma and nivolumab (Opdivo) for treatment of advanced NSCLC.

Pembrolizumab and ipilimumab are targeted therapies known as immune checkpoint inhibitors. Both agents were designed to harness the body's immune system to fight cancer. However, the drugs have different molecular targets and affect the immune response to cancer in different ways. Ipilimumab, a monoclonal antibody, binds to a protein called CTLA4, which is found on the surface of T cells. CTLA4 is a checkpoint protein that normally acts to keep immune responses in check to prevent overly strong responses that might damage normal cells as well as abnormal cells. The binding of ipilimumab to CTLA4 relieves this suppression. Ipilimumab was the first immune checkpoint inhibitor to be approved by the FDA (in 2011) to treat advanced melanoma (National Cancer Institute website).

Pembrolizumab, also a monoclonal antibody, binds to PD-1, another protein on T cells. When PD-1 is activated by binding to a protein that is produced by many tumor cells, the immune response is suppressed. Binding of pembrolizumab to PD-1 blocks activation of the PD-1 pathway, allowing the immune response to proceed (National Cancer Institute website).

Nivolumab, which was initially approved for the treatment of metastatic melanoma, is the first immunotherapy drug to be approved to treat lung cancer. It works by inhibiting a protein receptor called PD-1 on T cells, a type of immune cell (National Cancer Institute website).

Several clinical trials established the efficacy of these immunotherapies. Interim results from an international randomized phase III trial show that patients with advanced melanoma who received pembrolizumab (Keytruda) had longer progression-free survival and overall survival than those who received ipilimumab (Yervoy) and experienced fewer adverse effects. In September 2014, the FDA granted accelerated approval to pembrolizumab as a second-line therapy for advanced melanoma that has progressed during treatment with ipilimumab or BRAF inhibitors. The approval was based on results from a randomized phase IB trial, called KEYNOTE-001. As a condition of this accelerated approval, Merck was required to conduct a multicenter randomized trial to establish the superiority of pembrolizumab over standard therapy and to describe its clinical benefit. Antoni Ribas, MD, PhD, of the University of California, Los Angeles, was the first author of the study, which was sponsored by Merck Sharp & Dohme, the maker of pembrolizumab. The findings were published online in *The Lancet Oncology* on June 24, 2015 (National Cancer Institute website).

In addition, findings from an early-phase clinical trial may point to a biomarker that identifies patients with advanced NSCLC most likely to respond to the immunotherapy drug pembrolizumab (Keytruda). In the trial, patients whose tumors expressed high levels of the protein PD-L1 were more likely to experience substantial reductions in their tumors following treatment with pembrolizumab than patients whose tumors had lower PD-L1 expression, according to results from an early-phase clinical trial. These patients also lived longer without their cancer progressing. These trial results were presented at the American Association for Cancer Research (AACR) annual meeting and published in the *New England Journal of Medicine* (National Cancer Institute website).

On March 4, 2015, the FDA approved nivolumab (Opdivo) to treat patients with advanced NSCLC that has progressed during or after treatment with platinum-based chemotherapy. The FDA based the approval on findings from a randomized phase III trial that enrolled 272 patients with advanced NSCLC who were assigned to receive either nivolumab or the chemotherapy drug docetaxel. According to the FDA, participants who received nivolumab had a 41% reduction in the risk of death and lived an average 3.2 months longer than those who received docetaxel. Approximately 30% of patients treated with nivolumab were alive two years after beginning treatment compared with 13% of patients treated with docetaxel (National Cancer Institute website).

"The fact that we now have evidence of efficacy in the most common lethal tumor, non-small cell lung cancer, solidly cements immunotherapy as a critical tool in our fight against cancer," said James Gulley, MD, director of the Medical Oncology Service in the NCI Center for Cancer Research. "The rapid and durable responses seen with immune checkpoint inhibitors is a unique aspect of their clinical activity," Gulley added. Researchers are already testing combining immune checkpoint inhibitors with other therapeutic modalities "to further increase the proportion of patients who respond" (National Cancer Institute website).

Systems biology

Personalized cancer medicine relies on the optimization of treatment strategy for an individual patient using genetic information resulting in high efficacy and low toxicity. However, according to Tea Pemovska of the Finnish Institute for Molecular Medicine, the genotype-to-phenotype translation is not such a straightforward exercise, since the majority of cancers do not harbor druggable mutations, and meaningful response in patients such as the BRAF V600E mutations in melanoma are limited due to cell plasticity, tumor heterogeneity, and compensatory signals. Pemovska concludes that effective therapies remain difficult to predict solely

based on genetic information. Using acute myeloid leukemia (AML) and drug-resistant CML as case examples, Pemovska demonstrates how individualized systems cancer medicine can enhance cancer treatment through profiling of drug resistance and sensitivities of tumors.

Adult AML is the most common acute leukemia in adults and consists of a lineage of white blood cells in a heterogeneous clonal disease. It has poor survival, and the molecular mechanisms are currently poorly understood, alongside very few targeted therapeutic strategies. Yet, AML tumors are easy to sample and get large numbers of cells, with a high possibility to generate sequential samples. AML is also less genetically complex than typical solid tumors (The Cancer Genome Atlas Research Network, 2013).

According to Pemovska et al. (2013), the individualized systems approach to optimize therapies for leukemia patients is encapsulated in Figure 5.7.

The approach consists of assembling a concise and comprehensive functional screening collection, of which there are 161 active substances of approved small oncology substances, oncology-related approved substances, investigational compounds, and probes. Drug sensitivity and resistance testing would take place on these oncology patients with the compounds envisioned by

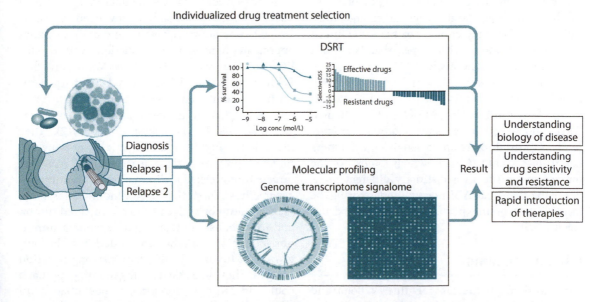

Figure 5.7 Drug sensitivity and resistance testing produces an understanding of the biology of the disease and the rapid introduction of therapies. DSRT, drug sensitivity and resistance testing; DSS, drug sensitivity score. (Reprinted with permission from Pemovska, T. et al. 2013. *Cancer Discovery* 3:1416–1429.)

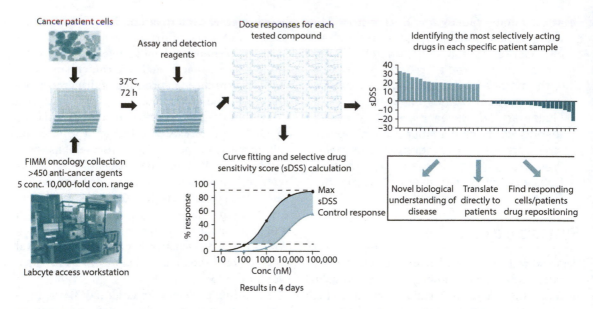

Figure 5.8 A drug sensitivity score is achieved for dose responses for each compound tested, identifying the most selectively acting drugs in each specific patient sample. FIMM, Finnish Institute for Molecular Medicine. (Reprinted with permission from Pemovska, T. et al. 2015. *Drug Discovery World*, 16: 47–53.)

the workflow shown in Figure 5.8 that would identify the mostly selectively acting drugs in each specific patient sample.

Drugs can be hierarchically classified and grouped into broad groups based on their response. Links between drug response profiles of drugs inhibiting different targets begin to emerge, indicating that they may connect to the same linked signaling pathways, with hopes that combined treatment will be synergistic.

In the case of AML, data by Pemovska's group revealed novel synergism between FLT3 inhibitors and dasatinib against M5 FLT3-ITD-mutated AML, and selectivity sensitivity of axitinib (currently approved for relapsed renal cell carcinoma for drug-resistant CML), which binds to the mutant-active site in a distinctly different manner than other BCR-ABL1 drugs, underscoring the identification of unexpected and new uses of drugs (Figure 5.9).

Cancer vaccines

Early work from the labs of Thierry Boon and Hans Schreiber (among others) suggested that tumor-specific mutations can sometimes function as tumor-specific antigens. James Allison and Bert Vogelstein predicted that many tumors should express mutational antigens based on their genomic repertoire of coding variants, and these might be the ideal tumor-specific targets for cancer immunotherapy. In the past, identifying the tumor mutation landscape and its most immunogenic peptides has been hampered by technical obstacles, most of which have now been overcome by next-generation sequencing and bioinformatics approaches to neoantigen prediction. A study published in *Nature* in 2012 by Elaine Mardis and Bob Schreiber's labs at Washington University demonstrated combined exome capture and *in silico* epitope prediction in a chemically induced mouse sarcoma model (Matsushita et al., 2012). They identified a highly immunogenic tumor-specific mutated protein antigen (spectrin BETA2) that targets tumor cells for elimination in an immune-capable host. First, they used genomic analysis to identify a tumor antigen from an unedited tumor, and to demonstrate that T-cell-dependent immunoselection is a mechanism underlying the outgrowth of tumor cells that lack a strong rejection antigen. They then asked if the genomics approach could identify tumor-specific neoantigens in edited, clinically apparent, and progressively growing tumors, and if the neoantigens identified in tumor rejection were the same as those recognized by checkpoint blockade immunotherapeutics.

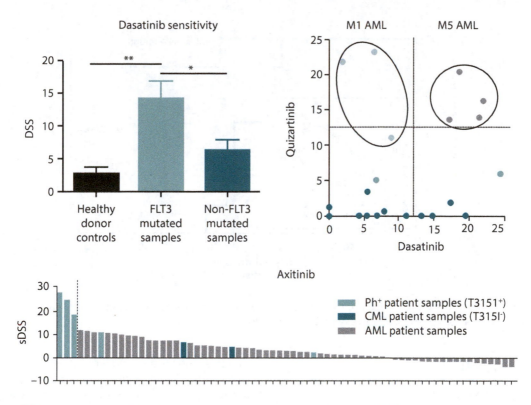

Figure 5.9 In the case of AML, data revealed novel synergism between FLT3 inhibitors and dasatinib against M5 FLT3-ITD mutated AML, and selectivity sensitivity of axitinib. (Reprinted with permission from Pemovska, T. et al. 2015. *Drug Discovery World*, 16:47–53.)

They also investigated whether tumor-specific neoantigens could be used to develop tumor-specific (personalized) cancer immunotherapies (Gubin et al., 2014). Data revealed tumor rejection after PD-1 immunotherapy. The study identified mutated class I epitopes by conducting cDNA Cap-Seq on tumor versus normal cells to identify expressed mutations, used multiple algorithms to predict epitopes, synthesized peptides, and then performed *in vivo* analyses and *in vitro* analysis to validate. These peptides were shown to protect against tumor challenge as a prophylactic long-peptide vaccine.

In addition, using the long-peptide neoantigen cancer vaccine in mice with established and growing sarcomas, the mouse survival increased and the mean tumor diameter shrunk versus human papillomavirus (HPV) controls (adapted from a study published by Matsushita et al. in *Nature* in 2012) (Figure 5.10).

In a recent study published in *Science*, Carreno et al. (2015) conducted genome-guided cancer immunotherapy in a first-in-human trial for metastatic melanoma patients. Dendritic cell vaccines that were conditioned with neoantigens identified using genomics for each patient were infused into five metastatic melanoma patients in an FDA-approved investigational new drug (IND) protocol (please note that the *Science* paper only described results for the first three patients). They are currently being monitored. Carreno et al. (2015) concluded:

Our first-in-human trial has demonstrated safety and a partial response of eliciting CD8+ T-cell memory for the tumor-unique neoantigens in three patients. Two patients have recently received their vaccines and are being monitored. We used HPLC fractionation and mass spectrometry to demonstrate predicted peptides were bound by HLA A*02:01 in patient bloods. The T cell repertoire in each patient was shown to be composed of diverse clonotypes post-vaccination.

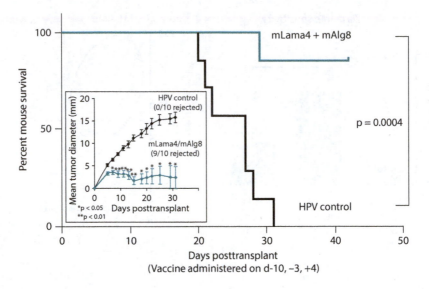

Figure 5.10 With the cancer vaccine, the mouse survival increased and the mean tumor diameter shrunk versus HPV controls. (Courtesy of Elaine Mardis.)

CONCLUSION

Kalia (2013, S12) writes:

For the treatment of cancer, the last decade has seen significant advances in personalized treatment as a result of two important scientific developments: (i) genomic analysis which has shown that common tumors such as breast cancer are, in fact, a mixture of several molecular entities; (ii) "targeted" drugs that inhibit specific biochemical pathways have become available. Advances in genomics and the application of genetic testing are now being used very effectively in oncology. This has a significant influence on cancer risk assessment, determination of prognosis, and selection of treatment. Clinical applications of novel genetic tools include: sequencing and analysis of germline genomic rearrangements of key cancer genes like BRCA1, BRCA2, and TP53; mismatched repair genes such as MLH1, MSH2, MSH6, and PMS2; development and widespread use of clinical karyotyping for hematologic malignancies; analysis of ERBB2 over expression in breast cancer; KRAS gene mutations in CRC, and gene expression

analysis in breast cancer as a form of molecular pathology. Use is now being made of new technologies (genomic, proteomic, single-cell analyses and high-throughput phenotypic assays) and powerful computational methods for delineating relevant biological networks underlying the cellular and molecular origins of cancer. The insight gained from oncology is now being adapted for new diagnostic and therapeutic strategies relating to other human diseases. For example, a new paradigm (P4) in systems medicine has evolved: "predictive, preventive, personalized and participatory (P4)" medicine which includes societal challenges, in addition to all the elements of oncology medicine. P4 medicine uses scientific, organizational and wellness strategies so that patients can access personalized medicine, thereby reducing the cost of healthcare.

According to Egalite et al. (2014), personalized oncology will bring new challenges to scientists and healthcare providers not seen previously in cancer care:

As stated in a position paper by the European Society of Medical Oncology

(ESMO), "A new era of personalized cancer medicine will touch every aspect of cancer care—from patient counseling, to cancer diagnosis, tumor classification, treatment and outcome—that demands a new level of in-depth education and collaboration between researchers, cancer specialists, patients and other stakeholders." This stated goal hides much complexity and there are considerable challenges, including ethical ones, that need to be addressed. Although many ethical issues have been identified and analyzed, whether in oncology or in genomics, some are still problematic because they are faced by healthcare providers and exacerbated by the use of newer sequencing techniques.

Ethical issues surrounding personalized oncology fall into four general categories: (1) informed consent, (2) privacy concerns, (3) return of results, and (4) cost of medicine and health disparities. Informed consent becomes problematic with the vast amount of information that must be conveyed to the patient to understand genetic testing, benefits, and risks. Due to ambiguity in the law concerning privacy and discrimination, patient concerns about disclosure remain. Integrating huge datasets of health information also exacerbates this situation. The decision on which results to return to patients has become a focus in the implementation of next-generation sequencing, especially surrounding the huge amounts of resulting data (Egalite et al., 2014). Counseling recommendations have been developed by the American College of Medical Genetics. Access to personalized cancer medicines raises ethical concerns about fair and just distribution of resources. Expensive targeted therapies may not always be covered entirely by insurance, leaving affluent patients paying for care out of pocket, while being out of reach to the poor.

However, despite these concerns, personalized oncology remains an improvement over the one-size-fits-all approach. Patricia LoRusso, DO, professor of medicine and associate director of innovative medicine at Yale Cancer Center, states that we are just starting to win the war on cancer. Three interdependent revolutions are taking place in cancer, LoRusso declares: a technology revolution in which the machinery of life is being created and changed, a systems biology revolution in which we are now beginning to understand the molecular basis of disease and mechanistic molecular strategies for therapeutics, and the individualized molecular revolution to tailor therapy to the patient and monitor and modify the therapy's efficacy and toxicity. Many members of LoRusso's family were affected by cancer in the seventies and had low survival rates. These same family members might be alive now, or at least would have had higher survival rates, with the current treatment regimes, which have been developed and exist to defeat their cancers, she concludes. On December 23, 1971, when President Nixon signed the war on cancer, there were no prepared minds or tools. With the recent approaches in genomics and immunology, a paradigm shift has occurred in personalized oncology and cancer care in general.

REFERENCES

Amado, R.G. et al. 2008. Wild-type KRAS is required for panitumumab efficacy in patients with metastatic colorectal cancer. *Journal of Clinical Oncology* 26:1626–1634.

Badalian-Very, G. 2014. Personalized medicine in hematology—A landmark from bench to bed. *Computational and Structural Biotechnology Journal* 10:70–77.

Bettegowda, C. et al. 2014. Detection of circulating tumor DNA in early- and late-stage human malignancies. *Science Translational Medicine* 6:224.

Blenkiron, C. et al. 2007. MicroRNA expression profiling of human breast cancer identifies new markers of tumor subtype. *Genome Biology* 8:R214.

Cagle, P.T., Raparia, K., and B.P. Portier. 2016. Emerging biomarkers in personalized therapy of lung cancer. In *Lung Cancer and Personalized Medicine: Novel Therapies and Clinical Management*, Ahmad, A., and Gadgeel, S.M., eds., 25–36. Gewerbestrasse, Switzerland: Springer International Publishing.

Calin, G.A. et al. 2002. Frequent deletions and down-regulation of micro-RNA genes miR15 and miR16 at 13q14 in chronic lymphocytic leukemia. *Proceedings of the National Academy of Sciences of the United States of America* 99:15524–15529.

Cancer Genome Atlas Research Network Collaborators. 2013. Genomic and epigenomic landscapes of adult de novo acute myeloid leukemia. *New England Journal of Medicine* 368:2059–2074.

Carreno, B.M. et al. 2015. Cancer immunotherapy. A dendritic cell vaccine increases the breadth and diversity of melanoma neoantigen-specific T cells. *Science* 348:803–808.

Chin, L., Andersen, J.N., and P.A. Futreal. 2011. Cancer genomics: From discovery science to personalized medicine. *Nature Medicine* 17:297–303.

Cho, S.-H., Jeon, J., and K. Seung II. 2012. Personalized medicine in breast cancer: A systematic review. *Journal of Breast Cancer* 15:265–272.

Dietel, M. et al. 2015. A 2015 update on predictive molecular pathology and its role in targeted cancer therapy: A review focusing on clinical relevance. *Cancer Gene Therapy* 22:417–430.

Egalite, N., Groisman, I.J., and G. Beatrice. 2014. Personalized medicine in oncology: Ethical implications for the delivery of healthcare. *Personalized Medicine* 11(7):659–668.

Ene, C.I., and E.C. Holland. 2015. Personalized medicine for gliomas. *Surgical Neurology International* 6(Suppl. 1):S89–S95.

Girotti, M.R. et al. 2016. Application of sequencing, liquid biopsies and patient-derived xenografts for personalized medicine in melanoma. *Cancer Discovery* 6:286–299.

Gubin, M.M. et al. 2014. Checkpoint blockade cancer immunotherapy targets tumour-specific mutant antigens. *Nature* 515:577–581.

Kalia, M. 2013. Personalized oncology: Recent advances and future challenges. *Metabolism Clinical and Experimental* 62:S11–S14.

Matsushita, H. et al. 2012. Cancer exome analysis reveals a T-cell-dependent mechanism of cancer immunoediting. *Nature* 482:400–404.

Michael, M.Z., O'Connor, S.M., van Holst Pellekaan, N.G., Young, G.P., and R.J. James. 2003. Reduced accumulation of specific microRNAs in colorectal neoplasia. *Molecular Cancer Research* 1:882–891.

National Cancer Institute website. http://www.cancer.gov.

Okimoto, R.A., and T.G. Bivona. 2014. Recent advances in personalized lung cancer medicine. *Personalized Medicine* 11:309–321.

Pemovska, T. et al. 2013. Individualized systems medicine strategy to tailor treatments for patients with chemorefractory acute myeloid leukemia. *Cancer Discovery* 3:1416–1429.

Pemovska, T., Ostling, P., Heckman, C., Kallioniemi, O., and W. Krister. 2015. Individualised systems medicine next-generation precision cancer medicine and drug positioning. *Drug Discovery World* 16:47–53.

Robinson, D. et al. 2015. Integrative clinical genomics of advanced prostate cancer. *Cell* 161:1215–1228.

Rodriguez-Antona, C., and M. Taron. 2015. Pharmacogenomic biomarkers for personalized cancer treatment. *Journal of Internal Medicine* 277:201–217.

Ryder, C.B., and C.L. Schmotzer. 2015. Circulating tumor DNA: The future of personalized medicine in oncology? *Clinical Chemistry* 61:443–444.

Takamizawa, J. et al. 2004. Reduced expression of the let-7 microRNAs in human lung cancers in association with shortened postoperative survival. *Cancer Research* 64:3753–3756.

The Cancer Genome Atlas Research Network. 2013. Genomic and epigenomic landscapes of adult de novo adult myeloid leukemia. *New England Journal of Medicine* 368(22):2059–2074.

Wijesinghe, P., and A. Bollig-Fischer. 2016. Lung cancer genomics in the era of accelerated targeted drug development. In *Lung Cancer and Personalized Medicine: Novel Therapies and Clinical Management*, Aamir, A., and Gadgeel, S.M., eds., 1–23. Gewerbestrasse, Switzerland: Springer International Publishing.

CHAPTER 6

Personalized medicine's impact on disease

How to move personalized medicine beyond oncology

McKinsey & Company (2013)

Many of the gains in personalized medicine have been in oncology. Oncology will continue to be the leading playgroup for personalized medicine, with a new understanding into disease and new diagnostic technologies (McKinsey & Company, 2013). Additionally, personalized medicine's impact extends beyond cancer medicine into diseases, such as AIDS, type II diabetes, central nervous system (CNS) disorders, prenatal disease, and immune-related, psychiatry, and cardiovascular diseases. However, these advances have been limited by genome-wide association studies (GWAS) that heralded ambiguous results and do not completely explain genetic heritability that would identify potential driver mutations and the difficulty in finding appropriate biomarkers. This chapter underscores the achievements in personalized medicine and pharmacogenetics for these categories of diseases.

TYPE II DIABETES

The impact of personalized medicine on type II diabetes has been underscored by limited GWAS that remain unable to predict diabetes mellitus type II (DM2) risk. Sixty-five gene variants explain only 10% of the heritability. Metabolomics has played a key role, with mass spectrometry identifying serum metabolites predicting DM risk and insulin resistance. The MTHFR gene has been shown to have an association for risk of diabetes complications. However, family history and clinical presentation remain the mainstays of diabetes care and prevention currently: "The principle of genome-wide association studies is to investigate differences in the prevalence of genetic variations (single nucleotide polymorphisms, SNPs) in DNA samples from populations with and without the condition of interest. Significant differences point to possible etiological associations with the condition" (Glauber et al., 2014, 3). According to Glauber et al. (2014), a large number of genes contribute to diabetes risk, including CDKAL1, CDKN2A, and CDKN2B, which influence β-cell mass; MTNR1B, TCF7L2, and KCNJ11, which influence β-cell function; FTO, which is associated with obesity; and IRS1 and PPAR-γ, which contribute to insulin resistance independent of obesity. Biomarkers for genetic susceptibility to diabetes in certain ethnic groups with a high incidence of diabetes and interactions between individual genetic variants may also influence DM risk. GWAS have identified 65 gene variants, but these variants only explain 10% of the heritability (Glauber et al., 2014).

Pharmacogenetics testing has determined the safety and efficacy of oral antidiabetic therapy such as metformin, but these advances have not yet reached the clinic. According to Glauber et al. (2014, 5), "Initial studies with a limited number of DNA markers showed only modest incremental value of adding genetic data to clinical information in predicting risk for Type II diabetes. Thus the potential for genomics to enhance prediction of DM2 risk remains unrealized."

Targeted metabolic studies have identified metabolites in the blood and urine that may predict DM risk and insulin resistance, including circulating levels of aromatic and branched-chain amino acids. "Using a targeted metabolomic approach and measuring over 160 serum metabolites with flow injection analysis tandem mass spectrometry in prospectively collected samples from large population-based studies, Floegel et al. identified a number of changes in sugar metabolites, amino acids, and choline-containing phospholipids that modestly improve prediction of DM risk. Identifying such metabolomic markers may prove to be useful in directing studies of the associated genes in at-risk populations" (Glauber et al., 6).

Advanced glycation end products and the gene for methylenetetrahydrofolate reductase (MTHFR) have shown association with risk for diabetes complications such as nephropathy and retinopathy.... Twenty genetic variants (SNPs) have been discovered for cardiovascular disease associated with diabetes resulting in such events as acute myocardial infarction, heart failure, stroke, sudden cardiac death. In patients with Type II diabetes, a genetic predisposition score derived from GWAS of Type II diabetes predisposition was independently associated with risk for cardiovascular complications, pointing to an overlapping etiological basis for Type II diabetes and cardiovascular disease. However, traditional clinical risk factor approaches may remain more efficacious in determining Type II diabetes risk (Glauber et al., 2014, 7–8).

RHEUMATOID ARTHRITIS

Rheumatoid arthritis constitutes an interesting case in personalized medicine wherein personalized intensive therapy with targeted therapies at early stages of disease has improved clinical outcomes, but the variability to clinical responsiveness remains unknown, and no definitive biomarkers exist. Anticitrullinated protein antibodies had the potential of resulting in biomarkers that would stratify patients, but it has not borne out. Treatment with disease-modifying antirheumatic drugs (DMARDs) based on joint destruction as evidenced by radiographic damage has improved prognosis; however, variability, including genetic variability, to clinical symptoms of erosive joint damage and inflammation, remains unknown. T. Huizinga (2015, 189, 179) from the Department of Rheumatology at the Leiden University Medical Center in the Netherlands writes in the *Journal of Internal Medicine*:

The approach of treating patients as early as possible and to a predefined level of activity, that is personalized treatment intensity, has led to considerable improvement in the outcome of patients with RA. The identification of two clearly distinct subsets of RA based on the presence or absence of anticitrullinated protein antibodies ACPAs has led to more homogenous groups of patients with respect to their clinical course but has not resulted in the identification of specific disease pathways in these two groups. Despite the recognition that the prognosis and pathogenesis of patients in these groups differ, the principle of treating RA should focus on reducing disease activity. Finally, no predictive markers of response have been identified for the current targeted therapies.... Success of personalized medicine in RA: treat to target. The major characteristics of RA are joint pain and swelling leading to disability.

Erosive joint damage and inflammation were determined in cohort studies to be associated with joint destruction, as shown by disease activity and radiographic damage. Treatment with DMARDs of patients with predefined levels of activity improved prognosis (Huizinga, 2015).

Recent studies showed that immediate onset of therapy and the achievement of low disease activity improve long-term outcomes in inflammatory arthritis. As a result, considerable research efforts have focused on identifying synovitis as early as possible. Early intervention (symptom duration of ≤3 months) with DMARDs, compared with treatment starting just a few months later, led to dramatic differences in overall disease activity and destructive impact.... Treatment with methotrexate (MTX) in patients with undifferentiated arthritis (UA) with a disease duration of several months appears to delay the evolution to full-blown RA.... Overall, the vast majority of evidence supports the conclusion that the main contribution to variability of clinical responsiveness to therapeutic agents in RA remains unknown (Huizinga, 2015, 179, 179, 183).

MULTIPLE SCLEROSIS

GWAS have identified risk alleles for multiple sclerosis (MS). In a GWAS of 10,000 individuals of European descent, a human leukocyte antigen (HLA) allelic variant was found to protect against MS. Natalizumab, a risk-based therapy that stratifies patients according to serum anti-JC antibody status, has been shown as personalized therapy for MS. Monoclonal antibody treatment that selectively targets and depletes the CD20+ B cells to suppress the inflammatory symptoms of MS has been shown to have high efficacy and fewer adverse drug reactions (Matthews, 2015). According to Paul Matthews, who writes in *Nature Reviews Neurology*:

The past decade has finally seen genetics come of age as an important tool for the molecular analysis of disease mechanisms in MS. The collaborative development of an MS genetic database of unprecedented size has enabled the International MS Genetics Consortium to make major discoveries about susceptibility factors. A genome wide association study that included almost 10,000 people of European descent with MS identified specific HLA-DRB1 risk alleles and confirmed a role for HLA-A allelic variants in protecting against MS. Such genetic studies have revealed that immunologically relevant genes (particularly those involved in T helper cell differentiation) are substantially over-represented among those associated with susceptibility to MS, offering further support to the hypothesis of a primary immunopathogenesis. Work in the next decade will address how the interactions of these genes with environmental factors (such as sunlight and vitamin D) contribute to variation in the severity of the disease. Nevertheless, defining the molecular mechanisms of MS remains a challenge as long as the disease is mainly diagnosed on the basis of characteristic features that exist in an appropriate context, rather than by the presence of aetiologically specific features or biomarkers....

Six new treatments for MS have been approved since 2005, however evidence of personalized treatment for MS was seen in the development of natalizumab, a risk-based therapy which stratifies patients according to serum anti-JC antibody status. A monitoring regimen involving regular measurements of this antibody and MRI surveillance and evaluation serve as biomarkers for this immunosuppressive therapy (Matthews, 2015).

CARDIOVASCULAR DISEASE

Personalized medicine has made great inroads in cardiovascular disease in the area of pharmacogenomics, whereas GWAS, in terms of prevention and clinical course, may have not revealed risk or susceptibilities. GWAS have identified risk alleles that are associated with nearby causative variants, but the effect of these SNPs on phenotype is unknown. Whole genome sequencing may hold more promise. Pharmacogenomics, on the other hand, for the cardiovascular drugs clopidogrel (Plavix) and warfarin has established dosing relationships for individuals harboring alleles for genes that render them poor or fast metabolizers, leading to Food and Drug Administration (FDA) warning labels. Clopidogrel is activated by CYP2C19; a loss-of-function mutation in this enzyme leads to major adverse cardiovascular events, while a gain-of-function mutation leads to increased bleeding.

Claude Lenfant (2013, S7), former director of the National Heart, Lung, and Blood Institute at the National Institutes of Health (NIH), writes

The aim of GWAS is to identify the association of genetic variants, most often SNPs (but also duplications or deletions of stretches of DNA and other sequence variants) with traits or diseases. In general, GWAS reveal that in fact many genetic variants modulate a phenotype but for coronary disease each has a modest effect associated with increased risk well below the 2–3 fold risk seen with common risk factors, such as smoking or hypertension. Since 2007, GWAS have identified genetic associations with numerous cardiovascular diseases: myocardial infarction, coronary artery disease, abdominal aortic and intracranial aneurysms, peripheral artery disease, heart failure, atrial fibrillation, and atherosclerosis. Over the years, the results of GWAS have been evaluated and have revealed or emphasized some very important facts.

According to Lenfant (2013, S7):

The risk alleles identified in GWAS are seldom the true causative variant but are likely to be associated with, or inherited with, the nearby causative variant. Many cardiovascular diseases, especially coronary artery disease and myocardial infarction, are associated with many genetic variants, in many different locations, or loci, scattered throughout the genome. Up to 30 loci have been associated with coronary artery disease and myocardial infarction. How the genetic variants affect phenotype is unknown.

Lenfant (2013, S7) offers caveats on the results of GWAS:

GWAS have been very useful to identify the large number of genetic variants associated with cardiovascular diseases. However, they have failed to explain the mechanisms of these associations, thus not predicting preclinical diagnosis or offering directions for possible prevention, risk stratification assessment, and specific therapeutic interventions— the goals of personalized medicine. Research pathways have been identified that will lead to personalized therapies, and the pace of research should accelerate as whole genome sequencing becomes more accessible with improved technology.

In other words, GWAS have not offered explanations for the causes of cardiovascular disease or any options for prevention or cures. Whole genome sequencing may provide alternate pathways. But Lenfant (2013, S7–S8) adds:

Cardiovascular pharmacogenetics is an active field, providing clinically valuable information regarding individualized, personalized drug prescriptions. Many drugs available for the management of cardiovascular diseases have been extensively studied, especially two of the most prescribed drugs worldwide: statins and clopidogrel, an antiplatelet agent.

Clopidogrel has also been the subject of extensive pharmacogenetic studies to try to uncover the molecular basis of its effectiveness, or lack thereof; it is administered as a prodrug that must be activated by CYP2C19, an enzyme responsible for the metabolism of many drugs. Several variant forms of CYP2C19 have been identified and found to alter the concentration of active metabolites of CYP2C19. Their effect may be a loss-of-function or a gain-of-function, the former reducing the level of active metabolite causing the carriers to have a diminished platelet inhibition and an increased rate of major adverse cardiovascular events. In contrast, the carriers of the gain-of-function variants experience a reduction in the rate of adverse cardiovascular events, but also the risk of increased bleeding. The frequency of the loss-of-function or gain-of-function varies between subjects of different ancestry; in Europeans, Africans, and East Asians, the frequency of loss-of-function homozygotes is 15%, 15% and 29% respectively, while the gain-of-function is 21%, 16% and 3%.

Variants for CYP2C19 vary among ethnicities.

Cardiovascular drugs such as warfarin and clopidogrel have been shown to have pharmacogenetics effects that warrant drug labels from the FDA. Lenfant adds that "it is noteworthy that the FDA requires clinical trials to establish improvement of patient outcomes and others recommend the development of evidence-based clinical practice guidelines" (Lenfant, 2013, S8).

Johnson and Cavallari (2013, 988), in *Pharmacological Reviews*, give a thorough and outstanding assessment of clopidogrel and warfarin pharmacogenetics:

> The body of literature that has led to the clinical implementation of CYP2C19 genotyping for clopidogrel, VKORC1, CYP2C9; and CYP4F2 for warfarin; and SLCO1B1 for statins is comprehensively described.... It is anticipated that genetic information will increasingly be available on patients, and it is important to identify those examples where the evidence is sufficiently robust and predictive to use genetic information to guide clinical decisions.

CLOPIDOGREL PHARMACOGENETICS

Clopidogrel is a widely prescribed drug for cardiovascular disease. The genetic variability

contributing to drug metabolism of clopidogrel was discovered several years after its approval by the FDA.

Clopidogrel is a thienopyridine antiplatelet drug used in patients after an acute coronary syndrome or percutaneous coronary intervention to prevent future cardiovascular events. It works by binding to the platelet purinergic P2Y12 receptor and irreversibly inhibiting adenosine diphosphate-mediated platelet activation and aggregation for the life of the platelet (~10 days). In the United States in 2011, clopidogrel was the number two selling drug among all prescription drugs by dollars ($6.8 billion) and number six by prescriptions (28 million), making it a major drug in the treatment of cardiovascular disease.

That genetic polymorphisms contribute to interpatient variability in drug metabolism via CYP2C19 was first recognized in 1994. That the common loss of function polymorphism (*2) for CYP2C19 was an important determinant of clopidogrel effect was not recognized until 2006, nearly a decade after its approval by the FDA. It was not until these and other data were published that the critical role of CYP2C19 in the bioactivation of clopidogrel was appreciated (Johnson and Cavallari, 2013, 993, 997).

CYP2C19 has a prevalent loss-of-function variant that has varying SNP frequencies among ethnic populations. One allele, *2, is a common loss-of-function variant, of which 60% of Asians carry at least one. Collectively, other types of polymorphisms lead to poor metabolizers (PMs) for CYP2C19 and intermediate metabolizers (IMs) among whites, blacks, and Asians. From these allele frequencies, it was concluded that a large majority of the population has impaired capacity to metabolize the drug, leading to an FDA warning label (Table 6.1).

CYP2C19 contains a common loss of function polymorphism, called *2, which creates a cryptic splice site and premature stop codon 20 amino acids later,

Table 6.1 FDA warning label on clopidogrel (Plavix)

Warning: Diminished effectiveness in poor metabolizers

- Effectiveness of Plavix depends on activation to an active metabolite by the cytochrome P450 (CYP) system, principally CYP2C19.
- Poor metabolizers treated with Plavix at recommended doses exhibit higher cardiovascular event rates after acute coronary syndrome (ACS) or percutaneous coronary intervention (PCI) than patients with normal CYP2C19 function.
- Tests are available to identify a patient's CYP2C19 genotype and can be used as an aid in determining therapeutic strategy.
- Consider alternative treatment or treatment strategies in patients identified as CYP2C19 poor metabolizers.

Source: Reprinted with permission from Johnson, J.A., and L.H. Cavallari. 2013. *Pharmacological Reviews*, 65:987–1009.

resulting in loss of function. Allele frequencies for this single nucleotide polymorphism (SNP) are approximately 0.12, 0.15, and up to 0.35 in those of European, African, and Asian ancestry, respectively. This means that 25%–30% of those of European and African ancestry and up to 60% of Asians carry at least one loss of function *2 allele, which significantly impacts their ability to metabolize drugs via the CYP2C19 enzyme. Other less frequent loss of function alleles include *3, *4, *5, *6, *7, and *8. Among these, *3 and *8 are the most frequent. Details of these and other genetic polymorphisms that influence metabolism via CYP2C19 are comprehensively reviewed in the PharmGKB very important pharmacogene CYP2C19 summary. Collectively, these polymorphisms result in approximately 2%–5% of European and African ancestry individuals and 15% of Asians being homozygous for a loss of function allele, categorized as poor metabolizers for CYP2C19, with an additional

approximately 25%–35% of whites and blacks and 45%–50% of Asians who are heterozygous for loss of function alleles or intermediate metabolizers. Thus a large portion of the population has impaired capacity to metabolize via CYP2C19, which can have clinically important implications for certain drugs, including clopidogrel…. [In] March 2010, after the publication of a variety of studies suggesting risk based on CYP2C19 genotype, the FDA issued a boxed warning indicating potential for reduced efficacy (increased adverse cardiovascular outcomes) based on CYP2C19 genotype. Boxed warnings are the highest level of warning in the FDA-approved product label and are typically used to draw special attention of clinicians to issues of serious concern for the drug. The text of the clopidogrel boxed warning is shown [in Table 6.1]. The warning is limited to "poor metabolizers" (i.e., patients homozygous for loss of function alleles, e.g., *2*2) and focuses on post-acute coronary syndrome and patients undergoing PCI (Johnson and Cavallari, 2013, 993, 997).

Warfarin pharmacogenetics

Warfarin is a widely prescribed anticoagulant that has remained very difficult to dose for patients due to variability, and has been a major cause of adverse drug reactions. CYP2C9 and VKORC1 are the major genes influencing warfarin pharmacogenetics. There are more than 35 CYP2C9 alleles that lead to lower-dose requirements to prevent overbleeding and overanticoagulation. Variants in VKORC1 lead to warfarin resistance, necessitating high doses. Studies have maintained that genotype-guided therapy remains superior to standard dosing. The FDA also issued a warning label for warfarin dosing (Table 6.2).

Even with the recent approval of newer agents, warfarin remains the mainstay therapy for oral anticoagulation, accounting for an estimated 1.6 million treatment visits and over $144 million in expenditures in the United States in

Table 6.2 FDA warning label on warfarin indicating starting dose according to VKORC1 and CYP2C9 genotypes

Recommended warfarin starting dose in mg/day according to *VKORC1* and *CYP2C9* genotypes per the FDA-approved warfarin labeling					
	CYP2C9				
VKORC1-1963	*1/*1	*1/*2	*1/*3 or *2/*2	*2/*3	*3/*3
GG	5–7	5–7	3–4	3–4	0.5–2
AG	5–7	3–4	3–4	0.5–2	0.5–2
AA	3–4	3–4	0.5–2	0.5–2	0.5–2

Source: Reprinted with permission from Johnson, J.A., and L.H. Cavallari. 2013. *Pharmacological Reviews* 65:987–1009.

the fourth quarter of 2011. Warfarin also remains one of the most challenging medications to manage despite over 60 years of experience with the drug. In fact, warfarin currently ranks as the leading drug-related cause of serious adverse events leading to hospitalization in the United States. Challenges with warfarin stem from its narrow therapeutic index, wide interpatient variability in the dose required to achieve optimal anticoagulation, and numerous drug and food interactions.

CYP2C9 and VKORC1, which encode for vitamin K epoxide reductase complex 1, are the major genes influencing warfarin pharmacokinetics and pharmacodynamics, respectively. The CYP2C9 enzyme metabolizes the more potent S-enantiomer of warfarin to the inactive 7-hydroxy warfarin protein, whereas VKORC1 is the target protein of warfarin, responsible for converting vitamin K epoxide to its reduced form, which is an essential cofactor in carboxylation and activation of clotting factors II, VII, IX, and X (Johnson and Cavallari, 2013, 999).

CYP2C9

There are over 35 CYP2C9 alleles, of which the CYP2C9*2 and *3 alleles are most extensively studied and result from nonsynonymous SNPs in the gene's coding region that are important for enzyme activity. S-warfarin clearance

is reduced by approximately 40% with CYPC9*2 and 75% with CYP2C9*3. Accordingly, warfarin dose requirements are approximately 20% lower with the CYP2C9*1/*2 genotype and 35% lower with the CYP2C9*1/*3 genotype compared with the CYP2C9*1/*1 genotype. Doses of 1 mg/d or lower may be necessary in patients with the CYP2C9*3/*3 genotype to prevent overanticoagulation and bleeding (Johnson and Cavallari, 2013, 999–1000).

VKORC1

Nonsynonymous variants in the VKORC1 coding region lead to warfarin resistance, where very high doses (e.g., 0.20 mg/day) are necessary to obtain therapeutic anticoagulation.... [Data from] two large comparative effectiveness studies are available to support genotype-guided warfarin therapy. In the first study, nearly 900 patients starting warfarin therapy were offered free CYP2C9 and VKORC1 genotyping, with results provided to their physician with an interpretive report, and outcomes were compared with those from 2688 historical controls. During the initial 6 months of warfarin therapy, patients who underwent genotyping had 31% fewer hospitalizations for any cause and 28% fewer hospitalizations for bleeding or thromboembolism compared with controls. More recently, the CoumaGen-II study compared genotype-guided warfarin dosing using one of two pharmaco-genetic dosing algorithms in 504 total patients to standard dosing in a parallel control group (n = 1911). Patients in the genotype-guided arm were rapidly gen-otyped for CYP2C9 and VKORC1 vari-ants, with results available to inform the first dose. The investigators reported that genotype-guided therapy, regard-less of the algorithm used, was superior to standard dosing…. There were also fewer serious adverse events, including hemorrhagic events, thromboembolic events, and death, at 3 months with genotype-guided therapy (Johnson and Cavallari, 2013, 1000, 1001).

According to Johnson and Cavallari (2013, 1001):

In August 2007, the FDA approved the addition of pharmacogenetic data to the warfarin labeling. The label states that lower doses "should be considered for patients with certain genetic variations in CYP2C9 and VKORC1 enzymes." The warfarin labeling was further revised in January 2010 to include a dosing table based on CYP2C9 and VKORC1 genotypes. [Table 6.2, adapted from Johnson and Cavallari] may be used as a quick guide for clinicians to dose war-farin when genotype is available, realiz-ing that clinical factors still need to be taken into account. Pharmacogenetic information is also included in the war-farin labeling approved by the PMDA [Pharmaceuticals and Medical Devices Agency] in Japan and by regulatory body in Taiwan.

AIDS

A recent genomic study of the Thai HIV-infected population revealed that pharmacogenetics test-ing and biomarker testing improved clinical out-comes (Bushyakanist et al., 2015). This study was performed to mitigate the level of adverse drug reactions in HIV patients taking antiretroviral drugs, which included skin rash. According to the study, the benefits of pharmacogenetics test-ing are either guiding the initial drug regimen or individualizing regimen, increasing efficacy, and simultaneously avoiding adverse drug reactions. The conclusion: Use of pharmacogenetics testing in HIV-infected Thai adults should be considered based on HLA-B genotyping and CYP2B6 geno-typing data.

During the study period, a total of 103 HIV-infected patients with a median age of 46 (range, 20–85) years were enrolled … one out of five patients who [had] undergone HLA-B*4001 genotyp-ing developed lipodystrophy after receiv-ing stavudine (four of them did not receive stavudine). One out of two patients who had positive HLA-B*3505 developed skin rash after receiving NVP (nevirapine) (the other one did not receive NVP). Among 51 patients in whom CYP2B6 genotyp-ing was performed for the management and/or prevention of EFV (efavirenz) toxicity, 39.2% were extensive metabo-lizers. The most common CYP2B6 haplo-types were CYP2B6*1/*1, CYP2B6*1/*4, and CYP2B6*1/*2 with heterozygous mutant g.21563C.T (CT). Of these, 51.0% were intermediate metabolizers. The most common CYP2B6 haplotypes were CYP2B6*1/*6, CYP2B6*1/*2 with heterozygous mutant g.18492T.C (TC), CYP2B6*1/*5 with homozygous mutant g.18492T.C (CC), CYP2B6*2/*4 with mutations, and CYP2B6*2/*6 with muta-tions. The other 9.8% were slow metab-olizers. Six out of 51 patients in whom CYP2B6 genotyping was performed had more than one mutation, five patients had two mutations, and one patient had three mutations (Bushyakanist et al., 2015, 166–167).

PSYCHIATRY

Personalized medicine in psychiatry is at an early stage. Most therapies are guided by clini-cal presentation, but psychopharmacology based on gene variations is becoming more prevalent, particularly genotyping patients for CYP2D6 for antidepressants and antipsychotics. Neural net-works are starting to dictate drug development such as functional MRI. Jorge Costa e Silva (2013,

S40) of the Brazilian Brain Institute writes in *Metabolism*:

> Psychiatric patients tend to exhibit significant interindividual variability in their responses to psychoactive drugs, as well as an irregular clinical course. For these (and other) reasons, increasing numbers of psychiatrists are turning to genotyping for help in selecting the psychopharmacologic agents best suited to an individual patient's distinctive metabolic characteristics and clinical presentation. Fortunately, routine genotyping is already available for gene variations that code for proteins involved in neurotransmission, and for drug-metabolizing enzymes involved in the elimination of many medications. Thus, genotyping-based personalized psychiatry is now in sight. Increasing numbers of clinically useful DNA microarrays are in the development stage, including a simplified procedure for genotyping patients for CYP2D6, which metabolizes a high proportion of the currently prescribed antidepressants and antipsychotics. It has been pointed out that psychiatric disease is rarely a consequence of an abnormality in a single gene, but reflects the perturbations of complex intracellular networks in the brain. Thus, analysis of functional neuronal networks is becoming an essential component of drug development strategies. The integrated use of technologies such as electroencephalography, magnetoencephalography, functional magnetic resonance imaging (fMRI), and diffusion tensor imaging (DTI), in combination with pharmacogenetics, promises to transform our understanding of the mechanisms of psychiatric disorders and their treatment. The concept of network medicine envisions a time to come when drugs will be used to target a neural network rather than simply components within the network. Personalized medicine in psychiatry is still at an early stage, but it has a very promising future.

GWAS have identified copy number variations and SNPs between disease groups and control groups; however, GWAS remain challenging in psychiatry considering the multifactorial non-Mendelian style of inheritance and heterogeneity of clinical presentation. According to Alhajji and Nemeroff (2015, 395–396) of the Department of Behavioral Sciences and Psychiatry of the University of Miami School of Medicine:

> Genome-wide association studies (GWAS) have contributed to advances in personalized medicine in psychiatry by identifying genetic variants of disease that may contribute to disease vulnerability. The results have been promising in bipolar disorder and schizophrenia but considerably less so in major depression. Linkage studies preceded GWAS approaches and obtained information from family members with and without the specified disease. Thus far, these studies have been very useful for disorders with single-gene Mendelian type of inheritance, for example, Huntington's disease, but much less so when analyzing psychiatric disorders because of the multifactorial non Mendelian-style disease inheritance and heterogeneity of disease manifestations. Candidate gene association studies assess specific alleles or genetic markers, for example, single nucleotide polymorphisms (SNPs) and copy number variations, and measure whether these occur more frequently in family based studies or case-control trials. It is the authors' view that in spite of controversy in this area, this approach has been shown to be more informative in psychiatry research than the other approaches described earlier.

However, they add:

> It is important to note that large-scale GWAS studies can successfully identify highly statistically significant differences in the frequency of certain genetic polymorphisms in the disease group compared with the control group. The absolute difference in the frequency

is almost always quite small; individually, these variations constitute a very small percentage of disease heritability (Alhajji and Nemeroff, 2015, 395–396).

MAJOR DEPRESSIVE DISORDER AND BIPOLAR DISORDER

Some of the statistics surrounding major depressive disorder and bipolar disorder are as follows, including prevalence and genetic component.

The lifetime prevalence rate of major depressive disorder (MDD) is approximately 11%–13% in men and almost twice that in women with the genetic component risk of 30%–40%. Approximately 10% of the population older than 12 years report taking antidepressants in the United States, although the rate of response is highly variable among patients. Bipolar disorder affects approximately 1%–3% of the population depending on the diagnostic criteria used. Numerous family studies have confirmed that bipolar disorder is highly familial, with the risk increasing to 10-fold in first degree family members of patients with bipolar disorder. Bipolar disorder is known to have a strong genetic heritability, which contributes to approximately 60%–85% of the risk (Alhajji and Nemeroff, 2015, 397).

Alterations in drug-metabolizing enzymes

The CYP450 enzyme system has remained a focus of study for metabolism of antidepressants. Polymorphisms leads to PMs, IMs, extensive metabolizers, and ultrarapid metabolizers (UMs). PMs develop side effects as a result of increased half-life. UMs have poor response to antidepressants due to their shortened half-life. However, evidence is still not compelling for routine screening.

One major focus of pharmacogenomics has been the hepatic cytochrome P450 enzyme (CYP450) system, particularly the polymorphic CYP450 2D6 isoenzyme, which is responsible for the metabolism of most antidepressants. This isoenzyme is responsible for metabolizing tricyclic antidepressants (TCAs), many selective serotonin reuptake inhibitors (SSRIs), serotonin-norepinephrine reuptake inhibitors, and antipsychotics. Multiple studies have scrutinized CYP450 2D6 polymorphisms as a predictor of tolerability and side effect burden of antidepressants. Patients with polymorphisms of this isoenzyme are classified into 4 categories: (1) poor (PM), (2) intermediate, (3) extensive, and (4) ultra rapid metabolizers (URM). CYP-2D6 PMs are at a higher risk of developing side effects because of slower metabolism, increased half-life, and increased bioavailability. In contrast, CYTP450 2D6 URMs are more likely to exhibit a poor response to therapy because of the shortened half-life and decreased bioavailability. These factors should be taken into consideration when patients show severe side effects at low doses or lack of efficacy at maximally approved doses. Despite these findings, evidence for routine screening for CYP450 genetic polymorphisms is still not compelling (Alhajji and Nemeroff, 2015, 397).

Predictors of treatment response

There is a pressing need to identify biomarkers, not only to aid in identifying at-risk groups and to aid in diagnosis if depression but to also predict treatment response. In order to be successful, this will require...a far better understanding of the pathophysiology of these disorders. It is of paramount importance to determine if there are 5, 10, 50, or 200 distinct endophenotypes of MDD. Nevertheless, progress is beginning to be made (Alhajji and Nemeroff, 2015, 397). Neuroimaging techniques, such as functional MRI, and symptomology such as psychomotor functioning and executive memory and attention have been shown to serve as biomarkers and used to predict optimized treatment in depression. Insula metabolism was associated with good response to cognitive behavioral therapy and poor response to escitalopram (Alhajji and Nemeroff, 2015).

Alhajji and Nemeroff conclude that:

Despite some promising advances in personalized medicine in psychiatry, it is still in its early phases of development. Further research is required in order to develop tools that will impact clinical practice.

PHARMACOGENETICS OF SCHIZOPHRENIA

The FDA has approved the use of CYP2D6 activity in antipsychotic dosing, offering recommendations to reduce or avoid a number of antipsychotics in individuals who are nonextensive metabolizers. According to Pouget et al. (2014)* of the Pharmacogenetics Research Clinic in Toronto, Canada, who write in *Dialogues in Clinical Neuroscience*:

The CYP2D6 genotype has been most extensively investigated in association with antipsychotic metabolism, as approximately 40% of antipsychotics are major substrates for CYP2D6. CYP2D6 poor metabolizers have higher plasma levels of (dose-corrected) haloperidol, perphenazine, thioridazine, aripiprazole, iloperidone, and risperidone following antipsychotic treatment. The FDA has approved the use of CYP2D6 enzyme activity in antipsychotic prescribing decisions, providing recommendations to reduce the dose or avoid prescribing perphenazine, pimozide, thioridazine, aripiprazole, clozapine, iloperidone, and risperidone in individuals known to be nonextensive metabolizers.

Despite studies indicating a lack of association between CYP2D6 and antipsychotic efficacy, the genotype is considered a predictor of antipsychotic clinical outcomes and is included in pharmacogenetics tests.

The CYP2D6 genotype is robustly associated with clearance of several antipsychotics (including haloperidol, thioridazine, aripiprazole, iloperidone, and perphenazine), and has also shown some association with antipsychotic-induced side effects. Despite the clear role of CYP2D6 genotype in influencing antipsychotic metabolism, most pharmacogenetic studies have not found a significant association between CYP2D6 and antipsychotic efficacy. This may be due to the lack of correlation between antipsychotic plasma levels and clinical response (i.e., a great variability in each patient's dose-response curve), or challenges in the methodological design of clinical studies evaluating psychosis improvement. Despite these challenges, CYP2D6 is considered a predictor of antipsychotic treatment outcomes, and is included in all currently available commercial pharmacogenetic tests. CYP1A2 is another important enzyme with respect to antipsychotic pharmacokinetics, as approximately 20% of antipsychotics are major substrates for this enzyme. Although CYP1A2 activity is inducible by environmental factors such as caffeine and smoking, genetic factors are thought to account for a large portion of variability in CYP1A2 activity in the healthy population (Pouget et al., 2014).*

Antipsychotic response

Genetic variation in known antipsychotic drug targets may contribute to variability in response among patients by

* Text reproduced with from: Pouget JG, Shams TA, Tiwari AK. Pharmacogenetics and outcome with antipsychotic drugs. *Dialogues Clin Neurosci.* 2014;16(4):555–566, with permission from the publishers, Association La Conférence Hippocrate. Copyright ©AICH 2014.

influencing antipsychotic binding to cell membrane receptors and downstream signaling. Identifying replicable genetic variants associated with antipsychotic response has been challenging due to a number of factors including the complexity of antipsychotic response, the lack of standardized outcome measures and thresholds for significant improvement, confounding by non-genetic factors (such as previous antipsychotic treatment, patient compliance, smoking, and concurrent medications), and low statistical power due to small sample sizes. Additionally, although many studies have included patients treated with different antipsychotics, it remains unclear whether pharmacogenetic associations are general or drug-specific. Despite these challenges, a number of interesting pharmacogenetic findings have emerged in antipsychotic response (Pouget et al., 2014).*

Antipsychotic-induced side effects

With an estimated noncompliance rate of 42%, encouraging patients to remain on their antipsychotic medication is a major challenge in the treatment of schizophrenia. One of the strongest predictors of noncompliance among first episode schizophrenia patients is whether they experience harmful side effects. The identification of genetic predictors of antipsychotic-induced side effects holds the potential to provide a rational basis for treatment selection in a way that minimizes their occurrence, thereby improving patient compliance and long-term clinical outcomes. With this goal of predicting antipsychotic-induced side effects in mind, a number of studies have explored the association between genetic variants and serious side effects of antipsychotics, with greatest focus on weight gain, tardive dyskinesia, and agranulocytosis. Findings in this area have been generally more robust than for antipsychotic response, in terms of effect sizes and reported replication in independent samples. This may be a result of the more objective nature of adverse drug reactions, in contrast to the previously discussed complexities of defining antipsychotic response.

Weight gain is a common and serious side effect of antipsychotic treatment, with up to 30% of patients gaining ≥7% of their baseline weight. There is robust evidence that variation in genes coding for the serotonin 2C (HTR2C) and melanocortin 4 (MC4R) receptors are associated with antipsychotic-induced weight gain, with moderate-to-large effect sizes. The protein products of these genes play important roles in appetite regulation, and may present an opportunity for therapeutic development to prevent antipsychotic- induced weight gain in the future (Pouget et al., 2014).*

CONCLUSION

While GWAS have not been definitive and studies have not identified targeted therapies and risk alleles associated with disease mechanism and cause, gains have been made in cardiovascular disease and type II diabetes. Some targeted therapies, particularly for MS, have emerged. Pharmacogenomics has provided valuable insight into genetic variability in populations and treatment response and offered many options for dosing and avoiding adverse drug reactions. Additional clinical trials and studies will be conducted to add to the armamentarium of personalized approaches for these illnesses, giving clinicians more treatments to prescribe and patients more options, and hopefully through research and translational medicine, the future will realize this potential.

REFERENCES

Alhajji, L., and C.B. Nemeroff. 2015. Personalized medicine and mood disorders. *Psychiatric Clinics of North America* 38:395–403.

Bushyakanist, A., Puangpetch, A., Sukasem, C., and S. Kiertiburanakul. 2015. The use of pharmacogenetics in clinical practice for the treatment of individuals with HIV infection in Thailand. *Pharmacogenomics and Personalized Medicine* 8:163–170.

Glauber, H.S., Rishe, N., and E. Karnieli. 2014. Introduction to personalized medicine in diabetes mellitus. *Rambam Maimonides Medical Journal* 5:1–16.

Huizinga, T.W.J. 2015. Personalized medicine in rheumatoid arthritis: Is the glass half full or half empty? *Journal of Internal Medicine* 277:178–187.

Johnson, J.A., and L.H. Cavallari. 2013. Pharmacogenetics and cardiovascular disease—Implications for personalized medicine. *Pharmacological Reviews* 65:987–1009.

Lenfant, C. 2013. Prospects of personalized medicine in cardiovascular diseases. *Metabolism: Clinical and Experimental* 62:S6–S10.

Matthews, P.M. 2015. New drugs and personalized medicine for multiple sclerosis. *Nature Reviews Neurology* 11:614–616.

McKinsey & Company. 2013. *Personalized medicine: The path forward.* Washington, DC: McKinsey & Company. Available from http://www.mckinsey.com/~/media/McKinsey/dotcom/client_service/Pharma%20and%20Medical%20Products/PMP%20NEW/PDFs/McKinsey%20on%20Personalized%20Medicine%20March%202013.ashx.

Pouget, J.G., Shams, T.A., Tiwari, A.K., and D.J. Müller. 2014. Pharmacogenetics and outcome with antipsychotic drugs. *Dialogues in Clinical Neuroscience* 16:555–566.

Costa e Silva, J.A. 2013. Personalized medicine in psychiatry: New technologies and approaches. *Metabolism: Clinical and Experimental* 62:S40–S44.

CHAPTER 7

The genome in the clinic: Diagnosis, treatment, and education

A recent cover story in *TIME* magazine (December 24, 2012) asked, can we know our genome and still do anything about the genetic information we receive? There are indeed challenges to whole genome sequencing in the clinic in the era of clinical sequencing and personalized medicine. There is the issue of informed consent: Will patients consent to having their genomes sequenced? Where will the resulting terabytes of data be stored? Do individuals have the right to see their sequencing? And finally, how will the clinical significance of this sequencing data be interpreted and ultimately translated to an electronic medical health record?

We are arriving at an inflection point in personalized medicine, according to many observers in healthcare. How will society embrace and need personalized medicine? According to Ralph Snyderman, MD, chancellor emeritus of Duke University and former professor at the Duke University School of Medicine, currently healthcare delivery is not heading in the right direction for society to embrace personalized medicine in the next 10 years; however, major technology breakthroughs could dramatically change quality and cost-effectiveness of care through personalized medicine. While personalized medicine has had some specific areas of impact, it still has not really changed how medicine is practiced currently.

However, it should have major impact in the future. There exists tremendous pressure to decrease healthcare costs. The 2010 Patient Protection and Affordable Care Act (ACA) (which may be in the process of being repealed) is in the process of developing new models of reimbursement. More than anything else, the ACA is driving change in how healthcare is delivered in that we are moving from fee-for-service to shared savings as an inducement to deal with clinical costs. Bundling a variety of services into one package to pay for single episodes of care is the way of the future in healthcare payments.

Why will healthcare move toward personalized medicine? One reason is to improve the prevention of preventable diseases and reduce the impact of those diseases. Health constitutes more than the absence of disease; one can enhance one's health due to an inherited genome that also absorbs all the external factors in the environment one experiences. Disease is mostly subclinical: today when disease becomes clinically manifest and detectable, it is usually irreversible, leading to higher costs for the healthcare system. In the current healthcare paradigm, we are waiting until disease occurs and then treating it, and spending a lot of money to do so.

However, personalized medicine may in fact change this paradigm by recognizing basic facts of health and disease. With personalized medicine, physicians can quantify baseline risk, monitor progression, refine risk, monitor and define disease, and thus personalize therapy. Yet, transferring this paradigm into the way providers care for patients presents a challenge. Snyderman states that unless you have a practice that readily adapts to this paradigm, providers will have difficulty embracing this approach.

Snyderman adds that the applications of personalized medicine to improve the delivery of healthcare exist. While the tools and applications of personalized medicine are of limited value and difficult to market unless healthcare embraces them, we can nevertheless move from the current form of healthcare. The "medicine of the future" consists of a very coordinated personalized health plan for the individual. Decisions can be made with an internal heuristic informed by personalized medicine. If a personalized medicine test is better, physicians should know why, when, and how to use it.

COMPANION DIAGNOSTICS: "OMICS" AND THE TRANSLATION OF BIOMARKER DISCOVERIES INTO ROUTINE DIAGNOSTICS

Joshua Cohen, PhD, senior research fellow and research associate professor at the Tufts Center for the Study of Drug Development, has done extensive research on the role of companion diagnostics and personalized medicine and offers a discussion on it in advancing the efficiency of personalized medicine. Cohen explains an idealized state whereby companion diagnostics and personalized drugs are jointly developed:

> There are only seven codeveloped companion diagnostic and drug combinations. There are definitely dozens of *post hoc* companion diagnostics which are developed long after the drug has been in use, warfarin being a great example.

Cohen emphasizes that it in is the codevelopment of companion diagnostics and therapy where the real advancement of personalized medicine lies.

> In terms of codevelopment, that's where the disappointment comes in. The reality is very different from the hype surrounding personalized medicine. That could change. We are seeing more codevelopment than we are seeing before, and if we looked at the pipeline, you see that there are many more instances of codevelopment of therapeutic and diagnostic in the pipeline.

Whether or not that would be successful is a major question, which we can only speculate in terms of success rate. But the fact that there are in the pipelines, that in future years we will see more therapeutic diagnostic combinations and that is the ideal. When I say *ideal*, you want to be able to stratify the population not after the fact, you want to stratify during drug development process, not only will it be efficient, it will save money, because you don't want to try out these drugs which have terrible side effects on patients who will not benefit from them in the experimental phase. But you will also know in tandem that you can preselect the population that stands the most to gain from the drug benefit question. You will have a much better basis of evidence for any kind for any postmarketing approval or review of the drug.

Cohen cites Herceptin as a great example of a drug that has benefitted enormously from the stratification of the breast cancer population.

> For example, if you were to look at Herceptin, which has been around a long time, and to look at the cost-effectiveness studies on Herceptin with or without the diagnostic, you will see huge differences. Herceptin is a terribly cost-ineffective drug for those who are not Her-2 positive. In fact, it is pretty much worthless. So if you were not to stratify breast cancer patient population, your cost-effectiveness for Herceptin would be very poor. But if you do stratify, cost-effectiveness numbers are quite good. There's a benefit at the drug development stage and benefit at the postdevelopment stage to have codevelopment and stratification because at each of these stages the goals of personalized medicine are achieved. We are seeing this in oncology, psychiatry, multiple sclerosis, and cardiovascular disease.

The vision of personalized medicine lies at the end of the trial-and-error period.

Right now if you look at multiple sclerosis drugs, they don't do as well with their clinical effectiveness numbers. But a better stratification of population of those would benefit [from drug versus those who wouldn't] would make it a win–win situation. We would be giving drugs with less likelihood of side effects. [There would be] no more trial and error and reiteration. Right now, we take a drug that appears to be better than placebo in randomized clinical trial: it either works or doesn't work; it gives you side effects, or it doesn't give you side effects, and you went on to the next drug. Personalized medicine's hope, which is being realized in small steps, is that we won't be going through [an] unnecessary duration of [the] trial and error method. While a lot of science hasn't been uncovered concerning which biomarkers work, but if we were able to stratify populations across all therapeutic categories, we would have a much more efficient healthcare system. Efficiency is not just about cost; it's also about effectiveness. You would have more effectiveness for less money.

To achieve efficiency (effectiveness with cost savings), there have to exist a regulatory stimulus, reimbursement for diagnostics by payers, and partnerships between drug companies to codevelop companion diagnostics and personalized medicine drugs.

There are three points to consider in that road map to efficiency. There is the regulatory end. Regulatory guidance meant that FDA understood how to streamline a codevelopment of therapy and diagnostic.

Another thing, companion diagnostic, or any kind of diagnostic, reimbursement is antiquated. Methods that are out there date from the 1980s, but diagnostics today are far more complex than that. What does this mean? Payment schedules haven't been updated enough. In many instances, commercial payers, and Medicare/Medicaid, will underpay

for diagnostics or not pay [for it] at all, whereas they will pay for therapeutics and drugs. The reason they underpay is that they are paying for technical activity such as reading an assay, or lab activities, but they are not really paying for the value of the test itself. Herceptin and Gleevec, for example, are paid for by all payers, but diagnostics often or not. Payers are worried about the "Angelina Jolie effect," where people see a diagnosis in the news and will want to get tested for it. Payers are holding back.

Payers should pay for diagnostics that stratify populations.

However, payers are wrong not to pay for diagnostic which would stratify populations of those with breast cancer, chronic myelogenous leukemia, or lung cancer to likely responders versus nonlikely responders. It would be good for all involved, patients, providers, payers, even manufacturers to pay for these diagnostics, since they provide good information.

We should also keep in mind partnerships between diagnostic firms and drug companies. That needs to be bolstered, we see that to some extent, with Xalkori, which was approved in 2010. Pfizer developed the drug, and Abbott [Labs] developed the diagnostic; together they collaborated to create a great drug which should only be taken by [a] small subgroup of lung cancer patients. It is highly effective in that group of patients. We don't see enough of these partnerships. There are lots of diagnostic companies; however, they fall by the wayside because it is a tough business. This would improve the science of personalized medicine but also the marketing of the products. Getting it through approval process would involve more collaboration. In some instances, companies are wary of this, because they don't want to give their competitors too much of an advantage. However, because of the unique capabilities of each component, because the diagnostic firm knows something that the drug company doesn't [they do something that is different from the drug firm], their collaboration from phases I through III, right through postmarketing too, should occur more frequently.

However, as Cohen alluded to, discovery must advance also. The promise and the challenge of personalized healthcare lie in the concept of biomarkers as the foundation for companion diagnostic tests. Currently, 15 biomarkers have been translated into routine diagnostics. Since the list is relatively short, the issue becomes how this process can be accelerated. Cancer biomarkers have led to accurate risk assessment, screening, differential diagnosis, prognosis, and therapy response; however, to be clinically useful, they must lead to a change in clinician's judgment. The key tests must be highly reproducible for the clinical patient and the doctor to rely on the data. In terms of the implications and complications of companion diagnostic development, different kinds of variables impact analytic validity of the sample. For example, would there be manual or algorithmic interpretation? Walter Koch of Roche Molecular Systems states that 20% of current HER2 testing for breast cancer may be inaccurate, and 30% of laboratories are unable to report biomarkers such as the KRAS mutation.

There are additional elements to companion diagnostics. The emphasis in companion diagnostics has moved from finding single genes that cause diseases to finding sets of genes or groups of genes, RNAs, and proteins, which together can be found in association with disease or can provide targets for specific subgroups of disease. Researchers at companies, for example, Genomic Health, are seeking to find sets of genes that can predict subgroups within different diseases, such as cancers, and identify those that should be (and those that should not be) treated aggressively with chemotherapy.

In addition to the diagnoses of disease, the premise of companion diagnostics lies in a determination of which drug should be best used to treat the disease in a patient with consideration of a patient's results from companion diagnostics and targeting those therapies to be most effective. The goal is to focus on patients who would most likely benefit from certain targeted therapies. Thus, there is a huge opportunity for those companies that are trying to identify signatures and determine the predictive value of those signatures in treatment, and then to couple them with those treatments that show efficacy.

Consider DNA or a genetic diagnostic that could predict groups of patients that have slightly different versions of disease. For example, some groups of patients would benefit the most from a drug, some groups would benefit the most from exercise, and some groups that would benefit the most from changes in the physical environment.

Additionally, there could be RNA (DNA is transcribed into RNA in cells) profiles, and RNA profiles would necessitate a tissue that could be easily accessible. RNA profiles are often used in cancer since there is access to tumor cells. Moreover, there exist protein profiles, a third type of diagnostic. Also, there exist sets of metabolites that could be used as markers, known as metabolomics. Companion diagnostics constitute these sets of signature profiles to characterize disease.

To achieve codevelopment of drug and diagnostic, personalized medicine is not about trying to turn every patient into a different category; just finding the right drug for each and every patient would be virtually impossible and prohibitively expensive. This would not result in the economics we would hope to achieve, which is reducing treatment for patients, not wasting drugs on those patients, and limiting toxicity. From a stratification perspective, there are groups of patients who would not die or have serious side effects from adverse responses to drugs. They will be healthier and the result would be a better outcome because they are not taking drugs they cannot respond to.

A study published in *Cell* by Chen et al. (2012) demonstrated exactly this principle—that an integrative personal omics profile (iPOP) can be used for both monitoring health and diagnosing disease. This study group observed a healthy 54-year-old male over a 14-month period and analyzed genomic, transcriptomic, proteomic, and metabolomics information. The aim of the iPOP was threefold: (1) evaluate disease risk by analyzing an individual's genome, (2) study the expression of personal variants, and (3) use the omics data to gain a better understanding of the transition between physiological states. The study was very revealing. Not only did it demonstrate that molecular information can be used to estimate disease risk and for monitoring so that the disease can be treated, but it also painted a better picture of the dynamic processes occurring during the transition from healthy to diseased state.

Stephen Quake, PhD, who works in the Department of Bioengineering at Stanford, is pioneering new approaches to biologic measurement. In using single-molecule sequencing, his lab is proposing new measurement approaches that allow you to peer inside the body and measure health through a simple blood test. For example, to measure karyotype or aneuploidy (extra chromosomes in the fetus), the gold standard is amniocentesis. It is a pretty easy measurement, according to Quake; however, it is invasive. He suggests using circulating naked DNA that floats in blood. Some portion of DNA comes from the unborn baby; (a small percentage of DNA comes from the fetus), and thus there is a huge background. To solve the problem, use sequencers as molecular counters, looking for relative representations of chromosomes among molecules. Quake's lab conducted a study where investigators got cell-free DNA from pregnant women and did sequencing on that DNA to get a relative representation of DNA. He found a method for the perfect classification of Down's syndrome. We can ask this question of all chromosomes including "Trisomy 21," according to Quake, and a clinical study was conducted with successful identification of aneuploidy. This study was published in 2008; by 2013, 500,000 women had received his test.

Quake's applies the counting principle to the allele level to measure the fetal exome in order to discover aneuploidy and structural variation mutations in the fetus. This type of strategy was also applied in the clinic for heart transplants. A heart biopsy is invasive and expensive. But Quake asserts that a heart transplant is essentially a genome transplant. New heart cells die, there is then a mixture of cells in blood, the immune system attacks the heart, and then the level of cells derived from the heart rise in the blood. We can genotype the donor and recipient by sequencing cell-free DNA and asking what fraction of DNA is foreign single-nucleotide polymorphisms (SNPs). The foreign SNP load goes up with transplant rejection, as seen by a pathologist.

Quake asserts that this cell-free sequencing should in principle work with any solid organ transplant. His study has enrolled a larger cohort with large-scale validation than biopsy, which is the gold standard. This study has been published in the *Proceedings of the National Academy of Sciences* (Vlaminck et al., 2015).

EMERGING PERSONALIZED MEDICINE IN DISEASE: DIAGNOSTIC TESTS, BIOMARKERS, AND TARGETED TREATMENTS

Personalized medicine is impacting not only healthcare in theory, but also medicine in practice. A number of patients have improved clinical outcomes with personalized medicine tests and their associated biomarkers. Targeted treatments are now available based on genomic tests. Here, we look at five categories of disease for which personalized medicine is having an impact: cardiovascular disease, hypertension, mental illness, HIV, and cancer. Many companies in the business of personalized medicine, biomarker studies, and companion diagnostics are now making inroads in more effectively treating patients for these and other diseases. As a result, patients can avoid surgery and more drastic or simply ineffective treatments.

Cardiovascular disease

In the case of cardiovascular disease, more than $4.48 billion is spent every year treating symptoms of coronary artery disease (CAD). Many patients who arrive at the hospital or doctor's office with chest pain are sent for further testing and to a catheter lab; however, many of these patients are asymptomatic for CAD. The problem: Of the thousands of patients who are seen by a clinician for chest pain, only 10% have cardiac disease. A personalized medicine noninvasive test exists to help physicians decide which patients need to go on to further cardiovascular testing.

The market is large and costly for the cardiology division of a hospital to rule out the presence of cardiovascular disease. Some patients are directed to a catheter lab and undergo catheterization for more treatment. More than $4.4 billion is spent on these processes yearly. According to a study in the *New England Journal of Medicine*, only 40% of patients who undergo catheterization have cardiovascular disease (84% of patients received noninvasive tests) (Patel et al., 2010).

Corus's CardioDX is a multianalyte gene expression assay that was introduced as a less invasive way to identify obstructive CAD and assist primary care physicians and cardiologists in understanding whether the symptoms of cardiovascular disease in nondiabetic patients are caused by CAD. According to Mark Monane, the clinical diagnostic company CardioDx of Palo Alto, California, has devised a noninvasive cardiac test called Corus® to prevent physicians from treating patients with catheterization and further testing who do not have cardiovascular disease. Through findings that inflammation leads to atherosclerosis, scientists at CardioDx have devised a test that detects blood elements that serve as biomarkers for CAD. Among patients tested with the Corus CardioDx, those that receive a low score have been found clinically to have no cardiovascular disease. Thus, they likely do not need to undergo catheterization. The patient report rules out further symptomatic testing, and clinicians can figure out the alternative causes of patients' symptoms. The clinical utility of this test is high and clinical outcomes are robust, according to studies. As physicians are changing their course of treatment with the results from the Corus CAD test, the clinical validity is also high. In a six-month follow-up, no complications have been seen with the administration of the test.

Hypertension

The opposite of the personalization of medicine is standardization: to treat everyone the same. Many observers have commented, including advocates of personalized medicine, that the medical establishment is actually moving more deeply into standardization, rather than out toward personalization. One medical decision that could be personalized is the treatment of hypertension. Currently, in the case of hypertension, all patients are treated as the same target. According to Don Morris of Archimedes, Inc., a healthcare modeling company based in San Francisco, the fundamental barrier to personalization in the clinic is how medicine is currently practiced.

Morris states that "we are still using the same rules for medical decisions as in the precomputer age. We need to talk about a different paradigm (concerning) how (medical) decisions are made." His company, Archimedes, utilizes mathematical models to transform medicine.

The personalized medicine approach to treating patients with statins (used for treating hypertensive patients) determines who will and will not benefit from statins before prescribing them. Archimedes takes individual data from trials and studies and constructs very accurate mathematical models to determine patient benefit for taking statins. Through the use of 30 input variables, the result is very personalized and answers the crucial health questions of the clinician and patient: What is the risk of myocardial infarction, and how beneficial or harmful is the drug (in this case, the statin) to the patient? According to Morris, Archimedes IndiGO, a decision support tool already in use at Kaiser Permanente and other organizations, computes patient data to generate personalized guidelines. IndiGO's new risk-based decision paradigm can easily be extended, allowing for a seamless integration of a variety of complex information in the clinic.

The existing national guidelines for the treatment of hypertension are very different from personalized guidelines. In the Archimedes personalized medicine model, there exists the potential for increasing quality and reducing costs. The company has devised personalized risk profiles for patients where they can lower the risk of heart attack, stroke, and so forth, by taking a number of actions, including taking a statin, taking aspirin, stopping smoking, and taking blood pressure medication. The result: A patient can see how risk, in reliable, numerical terms, of morbidity gets reduced in a printout. The outcome according to Morris: The patient is engaged. Additionally, this approach is expandable to include any kind of clinical information, not just that for hypertension. Morris notes that six U.S. states and Kaiser are employing this model to reduce hypertension in patients.

Mental health

The key to understanding personalized medicine's impact on mental health is to realize that psychiatric illnesses are not diseases, but disorders. "DSM-5 [the diagnostic manual for psychiatry] lists clinical syndromes," according to Stephen Stahl, professor of psychiatry at the University of California, San Diego. Stahl explains that there has been a move from a diagnostic system based only on symptoms that a patient presents, to a psychiatrist one based on biosignatures and finding links between biomarkers and treatment responses.

Currently in the practice of psychiatry, the topographical orientation of information maps to symptoms. The DSM-5 currently indicates a symptom domain. However, one can move from clinical subtype to molecular abnormality. However, Stahl adds a necessary distinction: "Genes don't code for psychiatric disorders. Genes code for proteins. Proteins regulate the circuits associated with mental health disorders."

Potential targets for psychiatric drugs are the genes that regulate monoamines and can serve as biomarkers and plausible targets in the treatment of depression. An early marker is a natural product, L-methylfoltate, which is regulated by the enzyme methyltetrahydrafolate reductase (MTHFR). If L-methylfolate is low, depressive symptoms increase. The effect size of the L-methylfolate response rate is low, and there exists pharmacologic synergy with other psychotropic drugs. While this example is the only known one of psychiatry utilizing biomarkers, finding markers can be linked to the therapeutic efficacy of given agents, and there exists the potential for a companion diagnostic test that can be specifically associated with a specific therapeutic agent. Finding these biomarkers is critical when trial and error does not work to determine the right treatments for symptomatic psychiatric patients.

Thyroid cancer

Carol Berry of Asuragen, Inc. (a molecular diagnostics company in Austin, Texas), in discussing the molecular diagnostic–based treatment for thyroid cancer, states that a diagnostic test must meet a number of requirements before it can have clinical utility. First, does the test answer an unmet clinical need, and does it change the course of a physician's actions? And, along with its clinical utility, is the test accessible and easy to order?

In the case of thyroid disease, traditional cytology is 10%–40% indeterminate. Surgeons are doing too many second surgeries because they find thyroid cancer on the first surgery in patients. However, biomarkers in the form of mutations and translocations were discovered and account for 70%–80% of positive cancers. If there is no mutation, then the surgery is canceled. A thyroid molecular diagnostic test points the physician in the right direction and allows for the surgery to take place at the appropriate time.

Additionally, the study of microRNAs, small ribonucleotide molecules in the cell, can close the gap even further. The combination of mutations, translocations, and microRNAs constitutes multiple markers that, when put together, can significantly improve the diagnosis of thyroid cancer through preoperative molecular analysis of thyroid nodules.

Breast cancer

Steven Quay of Atossa Genetics, Inc., based in Seattle, and Joe Gray of the Oregon Health Sciences University discuss the personalized medicine dimensions for breast cancer. Breast cancer prevention through diagnosis and treatment may be advanced through targeting breast cancer lesions and understanding and translating the breast cancer genome.

For example, Atossa Genetics has four different personalized tests for breast cancer, and Quay asks the central question, what does one do with the molecular data? In noting that the origin of breast cancer is a terminal duct lobular unit, Quay states that breast cancer arises from obligatory precancerous atypical ductal hyperplasia (ADH). By targeting the presence, location, and cause of ADH, patients can be treated to prevent breast cancer with targeted therapy. One can locally identify the presence of ADH with a clinical test, and the cause of ADH with a laboratory test.

In a whole genome sequencing study, a precancerous duct in a woman was canulated and the transcriptome, or the series of genes transcribed in a cell, was examined for epithelial mesenchymal transition, a precursor to the development of breast cancer.

In looking for biomarkers, the breast cancer genome landscape consists of many copy number changes, or 45 regions of recurrent duplications or deletions. There exist ten distinct subtypes associated with outcome, and the spectrum of aberrations differs according to clinical subtype. In the breast cancer genome, in short, there are a few recurrent mutations, and many are rare.

Furthermore, individual tumors display substantial clonal heterogeneity, or diversity. Molecular diagnosis of breast cancer involves finding out which of the aberrations in individual tumors contribute to the deregulation of key pathways present in breast cancer, and finding out additionally how these abnormalities work together. The abnormalities also need to be mapped onto a phenotypic space, and algorithms need to be developed to integrate these abnormal pathways and determine which are upregulated or downregulated.

Dako's immunohistochemistry assay Hercep-Test™ is one of the brightest examples of the successful application of molecular diagnostics in oncology by targeting HER2/neu overexpression in metastatic breast cancer. The test enabled stratification of breast cancer patients into eligible and noneligible for Roche/Genentech's monoclonal antibody therapeutic Herceptin (trastuzumab) based on the prognostic biomarker HER2/neu presented in 20%–30% of breast cancer patients.

Colorectal cancer

Approximately 30%–50% of colorectal tumors are associated with an abnormal KRAS gene, signifying that nearly half of patients with colorectal cancer (CRC) might respond to anti-epidermal growth factor receptor (EGFR) treatment, and the other half might not. Two monoclonal antibodies, Erbitux (cetuximab) and Vectibix (panitumumab), have demonstrated favorable survival impact in populations with KRAS wild-type CRC. Hence, KRAS mutation testing is currently used in clinical practice to assist in identifying eligible patients for antibody treatment.

HIV

The HLA-B*5701 molecule is associated with hypersensitivity to the antiretroviral drug abacavir; hence, a positive test allows physicians to use the drug in a much larger population due to the assay's ability to identify patients that are at risk of developing severe adverse events.

IMPLEMENTATION OF PERSONALIZED DIAGNOSIS AND TREATMENT: EDUCATING PATIENTS AND HEALTHCARE PROVIDERS

The adoption of personalized medicine initiatives requires educating patients on the benefits of undergoing biomarker testing, finding targeting treatments, and training physicians in the fundamentals of genomic medicine. The administration of personalized medicine tests and their ultimate success in the clinic are contingent upon patient and provider awareness of the clinical validity and utility of a genomic (in this case, *personalized medicine* is used to mean *genomic*) test (especially when compared with a routine laboratory test) and the usefulness of targeted treatment. When and how should a physician use a biomarker test, and what decisions should she make based on it? While issues of payer reimbursement of companion diagnostics remain, a clear concern exists that people may not understand the impact of personalized medicine and its accompanying tests on medicine. This remains true of any disruptive innovation, and personalized medicine is no exception.

Even further, personalized medicine may in fact integrate into routine care, but what about its role in disease prevention? For example, many parents do not immunize their children; how many will even want their children's genome sequenced? One of the bottom-line issues for personalized medicine is public education: how educated does the public need to be before they will support these personalized medicine approaches? There also exist social and ethical issues: what are the risks to patients (young and old), the impediments to equity (per age, geography, education, socioeconomic status, religion, etc.), and overall implications for the existing clinical decision-making process (move toward patients and move toward doctors as complexity increases)? Furthermore, are there eugenic elements present?

Many of these topics remain the focus of the chapters that follow. However, I argue here that educating the public on personalized medicine can occur through initiatives backed by major medical centers and hospitals, including the many centers for personalized medicine and genomics that are emerging in academic institutions; government agencies; and industries, particularly the molecular diagnostic industry. For example, HMOs could hold seminars for their patients on personalized medicine and genomic tests and how they would benefit from them. Government agencies such as the National Institutes of Health could sponsor funding for grants that promote public education on personalized medicine. There could even emerge a government agency devoted to personalized medicine and education through Health and Human Services. Industry speakers could hold public workshops on genomics and personalized medicine and their implications.

Additionally, the process is under way for genomics education to be integrated into medical school training. The genomic test may be the future of healthcare, and physicians may have to work with teams of geneticists and genetic counselors, and whole genome sequence interpreters, on a routine basis, and it may be incumbent upon providers to understand the clinical significance of molecular data and the use of personalized diagnostics and treatment. In short, as personalized medicine becomes incorporated into the healthcare setting, increased awareness through the promotion of education initiatives must be made central.

Lamb and Gunter have encapsulated a chart of the needs for genetic education for clinicians, as shown below:

These skills focus not on assembling a set of genomic facts, but on creating a genomic context for thinking about health and disease.

- **Understanding modes of inheritance and the role of family history.** Common disorders, such as diabetes, schizophrenia, and many cancers, primarily occur due to multifactorial inheritance. However, rare Mendelian forms of these complex disorders do occur, and physicians should recognize their existence through the use of a three-generation family history. Family history is an important risk factor for common chronic diseases, and a well-documented family history is a useful tool for a primary care physician.
- **Identifying the appropriate genetic test to order for a patient and correctly interpreting its results.** Genetic testing, historically used to diagnose Mendelian disorders, is now being utilized for complex disease. Physicians need to understand the indications for a genetic test and which genetic variants should be tested. This is especially important given scientific and media reports about potential genetic associations with disease, many of which are not carefully communicated. Practitioners should also be versed regarding what a given genetic test is and is not analyzing, whether the results actually can guide patient care, and what the benefits and limitations of the genetic test are (including the implications of identifying a genetic change with unknown pathogenicity and whether the test has been evaluated in the correct population with respect to the patient).
- **Determining when individuals should be referred to genetic specialists based on calculation or estimation of genetic risk.** Healthcare providers need to recognize those "red flags" that indicate a patient is at a significantly increased risk for a disorder and understand when referral for detailed genetic or genomic services is appropriate.
- **Clearly communicating genetic information to patients for decision-making purposes.** Physicians should use language free of jargon, delivered at an appropriate level, to communicate risk, diagnosis, or treatment-related information to their patients. Often this involves referring the patient to appropriate websites, or support or advocacy groups for additional content (see section on *Continuing Education* for lists of these). Given the growth of the DTC industry and the speed in which information is distributed across the Internet and through social media sites such as Facebook and Twitter, in some cases patients arrive at the physician's office with a notebook of information on genetic factors gathered from sources with varying reliability. Sifting through what the patient has heard or read, and distinguishing truth from hype, is a critical skill for today's physician.
- **Being aware of and appropriately handling family dynamics.** Because genetics often involves risk transmission across generations, physician–patient interactions at times include gathering information from or sharing results with other members of the patient's family. Balancing a desire for privacy versus sharing information between family members may lead to ethical challenges.

Source: Reprinted with permission from Lamb, N.E., and C. Gunter. 2013. *Genomic and Personalized Medicine*, Ginsburg, G., and Willard, H., eds., 415–421. 2nd ed. Waltham, MA: Academic Press.

REFERENCES

Chen, R. et al. 2012. Personal omics profiling reveals dynamic molecular and medical phenotypes. *Cell* 48:1293–1307.

Lamb, N.E., and C. Gunter. 2013. Educational issues and strategies for genomic medicine. In *Genomic and Personalized Medicine*, Ginsburg, G., and Willard, H., eds., 415–421. 2nd ed. Waltham, MA: Academic Press.

Patel, M.R., Peterson, E.D., Dai, D., Brennan, M., Redberg, R.F., Anderson, V., Brindis, R.G., and P.S. Douglas. 2010. Low diagnostic yield of elective coronary angiography. *New England Journal of Medicine* 362:886–895.

Vlaminck, I.D. et al. 2015. Noninvasive monitoring of infection and rejection after lung transplantation. *Proceedings of the National Academy of Sciences* 1112(43):13336–13341.

CHAPTER 8

A new set of clinical tools for physicians

Medical centers around the country continue to see individuals and families who manifest medical disorders that defy diagnosis and do not fit with the current known disease. Similar to what the Beery twins went through, this lack of a diagnosis leads people on a diagnostic odyssey. Next-generation sequencing, particularly whole exome sequencing (WES), is providing a new set of diagnostic tools for physicians to provide patients with answers for ailing and perplexing medical problems. Additionally, there is an entire genomic healthcare team that use whole genome sequencing (WGS) and WES to help patients (some medical centers, such as the Medical College of Wisconsin in Milwaukee, the Mayo Clinic in Rochester, Minnesota, and Baylor University in Waco, Texas, have genomic healthcare teams).

Consider the case of pediatric patient Nicholas Volker, whose medical story was featured in a Pulitzer Prize-winning three-part series of articles in the *Milwaukee Journal Sentinel* in December 2010 entitled "One in a Billion: A Boy's Life, a Medical Mystery." Howard Jacob, PhD, of the Medical College of Wisconsin, the geneticist who sequenced Nic Volker, gave a firsthand account of the medical journey that Nicholas Volker had to endure. For Volker's rare, undiagnosed disease, DNA sequencing served as the confirmatory last resort of choice.

In July 2009, six-year-old Nic Volker was near death, suffering from inflammatory bowel disease and weight loss and at risk. Jacob sequenced Nic's DNA and found the identifying mutation that was causing the disease. It was one specific variant: xq25. Nic also had lymphoproliferative disease. The question remained: Should doctors do a bone marrow transplant on Nic because of his lymphoproliferative disease, and would this fix the gut disease? Nic was dying; maybe there was a good reason to do a bone marrow transplant. An umbilical cord blood transplant was done, which ultimately reconstituted the immune system and also affected the gastrointestinal system. Doctors still remained puzzled as to why this was the case for Nic, as to how his immune system affected the gut. Jacob concurred that genomics is a toolkit to provide better medicine. The issue is to figure out how variants cause disease. Is the variant real, and does the frequency of the variant equal the frequency of the disease. Which variants are actionable? Geneticists read a patient's DNA and find out what's wrong with the patient, who then undergoes a clinical procedure. However, why test if clinicians cannot do anything about it? However, Jacob notes that clinicians do this on a routine basis: for every diagnosis, such as hypertension, cancer, diabetes, and obesity, clinicians have a limited ability to affect health outcomes, and genomic medicine does not change that.

PRECISION MEDICINE IN COMMUNITY HEALTHCARE SETTINGS

Barbara McAneny, MD, CEO of New Mexico Oncology, discusses the challenges for a practicing physician incorporating precision medicine. She states that if you are going to have a precision medicine community, a practicing oncologist has a lot of information coming at him or her, sees 25–30 patients a day, and may not even have time to read a paper. Many of these patients are scared for their prognosis and may not be amenable to getting screened. It takes two weeks for a genetic test, and these patients want something now. A physician has to face which company to send the tissue to, and many of the tests do not correlate. It costs $5000 to get a whole panel of genes, with the insurance denying the claim and nothing revealed. The physician also has to reassure the patient not to become alarmed about every mutation if there are results. McAneny claims that cancer patients are twice as likely to declare bankruptcy. There are research system failures where finding mutations does not lead to enrollment in clinical trials.

This chapter examines the potential (and limits) of WES in ending patients' long and arduous journey, known as the diagnostic odyssey, by describing the nature of the genomic healthcare team, highlighting WES successes through case studies, and discussing the benefits and limitations of genome and exome sequencing for patients. The primary care provider, specialist, and surgeon will possess novel clinical tools in the "omics" era, in which personalized medicine will offer the necessary mechanisms to decipher intractable diagnostic dilemmas and possibly offer interventions to cure disease, as it did in the case of the Beery twins.

GENOMIC HEALTHCARE TEAM

There are various members of the healthcare team with specific responsibilities in providing individualized patient care. This section describes the criteria and processes by which patients receive care through WGS and WES.

Who will the patient meet, the primary physician, the physician specialist (geneticist or oncologist), or the genetic counselor? Who is behind the scenes? Laboratory professionals, bioinformaticians, and bioethicists. The primary physician can be from any discipline and recognizes the patient might benefit from genomic testing. The primary physician can be either a general practitioner or a specialist, and he or she refers the patient to the clinic, presents the case to the tumor board of the

hospital or clinic, and discusses the treatment plan after the results (when treatment exists).

The physician specialist can be an oncologist or a medical geneticist. The oncologist sees cancer patients prior to testing to ensure theirs is an appropriate case for genomic testing, and describes the process. The oncologist presents the case and results to the genomic tumor board and provides recommendations to the referring physician. The medical geneticist sees the patient to ensure all appropriate tests have been completed, and can also present the case and results to the tumor board, and may see the patient with the genetic counselor to provide final results and recommendations.

Genetic counselors are healthcare professionals who help people understand and adapt to the medical, psychological, and familial aspects of genetic contributions to disease. Their role is to provide pretest genetic counseling about testing risks, benefits, and limitations, and obtain informed consent. Genetic counselors authorize the submission of samples for WGS and WES (with insurance preauthorization). They also provide results and the case summary to the tumor or diagnostic odyssey board. Finally, they see the patient to explain the results and recommendations with a doctor when needed.

Genetic counselors have a master's in genetic counseling from a genetic counseling program that is accredited by the American Board of Genetic Counseling (ABGC). They are board certified in genetic counseling (CGC), and most enter the field from a variety of disciplines, including biology, genetics, nursing, psychology, public health, and social work. Genetic counseling is the process of helping people understand and adapt to the medical, psychological, and familial interpretations of genetic contributions to disease. Genetic counselors collect information of family and medical history, interpret family history, undertake risk assessment, and provide education on the disease, informed consent, and psychosocial support. With genomic counseling there is a higher likelihood for finding variants and incidental findings. Genomic counselors need more time to interpret the results.

There are numerous members of the genomic healthcare team who do not directly interface with the patient, including laboratory professionals, bioinformaticians, and bioethicists. Laboratory professionals are typically PhD scientists, including molecular pathologists and biologists, cytogeneticists, and biochemists, who help the team understand how individual genomic changes may or may not cause protein changes to affect a patients' disease. They also help the team define the meaning of genomic variants and whether any further clarifying tests are needed.

Bioinformaticians are a specialized field of professionals with computer science and biology backgrounds who analyze genomic data and filter and distill the trillions of nucleotides of genomic data into a smaller set of data that clinicians may find relevant to review for clinical purposes. Their training includes a MS or PhD in biological sciences and computational skills.

Bioethicists evaluate the ethical controversies brought about by advances in biology and medicine, and may have advanced degrees in a biomedical science, anthropology, sociology, philosophy, law, medicine, or nursing. Their role on an individualized medicine clinical team is to interface with national discussions on the ethical, legal, and social implications of genomics, and serve as the voice asking questions about respect, beneficence, and social justice. They ask the following: What are potential harms versus benefits (social, physical, and emotional) for the patient? How can we perform genomic testing to minimize harms and maximize benefit? They also study the clinic to understand patients' experiences, hopes, concerns, and expectations, and with their findings, they may recommend ways to improve patient treatment.

Personalized medicine clinical teams treat all cancer patients, including those with hematological (blood) malignancies and pediatric tumors, and diagnostic odyssey patients, who are undiagnosed patients with a suspected genetic component (mutation); a diagnosis may also provide comfort, resolve, or closure; information for family planning; and possible therapy. These are the types of patients who are considered candidates for genomic testing. The patient fit criteria are (Figures 8.1 and 8.2)

- Advanced cancer—have failed standard therapy
- Life expectancy greater than four months
- Able to withstand surgical excision
- Adequate tissue (quantity, quality, and accessible)

Tumor boards at personalized medicine cancer centers hold weekly meetings to review results, and include

- Oncologist (across sites)
- Genetic counselors (across sites)
- Clinic pathologists
- Laboratory directors
- Bioinformatics team representatives
- Bioethicist
- Disease expert (if necessary)
- Referring physician (if interested)

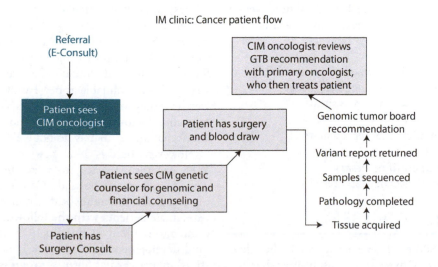

Figure 8.1 Personalized medicine cancer patient flow. After an external consult, the patient interfaces with the oncologist and receives genomic sequencing with variant reports returned. The genomic tumor board (GTB) then makes its recommendations.

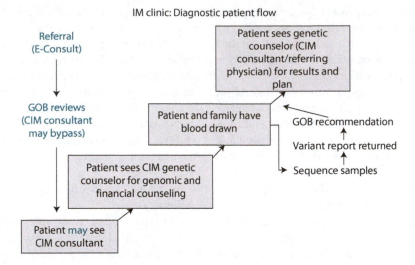

Figure 8.2 Personalized medicine clinic diagnostic patient flow. After a referral, the patient has a blood draw, and perhaps after seeing the personalized medicine oncology consultant, the patient has his or her samples sequenced, with variant reports returned. The GOB then provides recommendations and a plan based on the results.

Diagnostic odyssey patient fit criteria are

- Young disease onset
- Multiple organ system involvement
- May or may not have a family history of the same disorder and symptoms
- Multiple genetic tests have not revealed a diagnosis to date

Genomic odyssey boards (GOBs) hold weekly meetings and review cases and results, and include

- Center for Individualized Medicine (CIM) consultants or medical geneticists (across sites)
- Genetic counselors (across sites)
- Laboratory directors
- Bioinformatics team representatives
- Bioethicist
- Disease expert (if necessary)
- Referring physician (if interested)

DIAGNOSTIC ODYSSEY

Written in 800 BC, Homer's *Odyssey* is the tale of a Greek hero, Odysseus, on his 10-year, harrowing, and convoluted journey home after the fall of Troy. Today, an *odyssey* refers to a patient's long and eventful journey or experience in search of a

diagnosis. The patient seen on such a journey typically exhibits symptoms that just do not fit with current known diseases. The lack of a diagnosis leads people on diagnostic odysseys.

This section defines the role of the primary care provider in identifying patients who may benefit from genomic analysis, and discusses the benefits and limitations of genome and exome sequencing for diagnostic odyssey patients. This section also explains what patients need to know in order to consider genomic testing. Without a diagnosis, patients' families and healthcare providers may wonder if they are providing the right medical care for the affected individual: What are we missing? Are we losing valuable time? Families may also wonder if they would get a diagnosis if they saw just one more doctor, and often seek additional evaluations that unfortunately do not yield a diagnosis.

However, there is great news on the horizon: WES has the capacity to make diagnoses that could not previously be made. Consider the following case study. A diagnostic dilemma ensued when two female patients with the following symptoms came into the Mayo Clinic in 1999: short stature, mild developmental delay, thick bones, hard and stiff skin, pericardial fibrosis, progressive tracheal stenosis, and intense formation of scar tissue. All genetic testing was normal, and parents and siblings were unaffected.

Clinicians asked which gene might cause this. There was no way for the clinicians to gene hunt on two unrelated individuals unless they could think of a great candidate gene. The search continued. Eventually, the doctors, through their research, came across an illness known as Myhre syndrome—a clinical diagnosis reported 13 times since 1981—which looked as though it might apply to the two women. Myhre syndrome symptoms included mental subnormality, growth deficiency, muscular appearance, and in 9 out of 13 cases, stiff skin. None had tracheal stenosis, but similarity striking to these diagnostic dilemma cases. All 13 cases were sporadic.

In 2011 came a breakthrough: WES in Myhre syndrome revealed a *SMAD4* isoleucine 500 mutation in each person. The *SMAD4* mutation came as a great surprise to doctors. The *SMAD4* mutation is a known cause of hereditary hemorrhagic telangiectasia (HHT) (not seen in Myhre syndrome patients) and a known cause of juvenile polyposis (JP) syndrome (also not seen in Myhre syndrome patients). HHT and JP patients have no features of Myhre syndrome. Without genomics, the clinicians would not likely have considered *SMAD4* a candidate gene, thus demonstrating the power of genomics to surprise. Sequencing revealed that both individuals had the *SMAD4* mutations, which illustrated a new phenotype for a known gene.

There are too many examples of WES successes to list all of them here, which can be placed in different categories based on the types of genes discovered. WES is shown to be cost-effective and clinically useful. Sanger sequencing costs are about $1000 for the average gene. WES for very "genetically heterogeneous" disorders (i.e., caused by one of a potentially large number of genes), including familial amyotrophic lateral sclerosis (ALS), caused by 15 different genes; autosomal recessive deafness, caused by 39 genes; and Leigh's encephalopathy, caused by 35 genes. Studies of multiple family members with the same phenotypes led to the discovery of mutation in known genes, thus expanding their known phenotypes: familial leukemia, germline *p53* mutations found; fatal infantile encephalopathy, *SLC19A3*; autosomal dominant distal myopathy, *tropomyosin*; two members of a consanguineous family with macrocephaly and epiphyseal dysplasia, *KIF7*; and two relatives with osteoporosis, *CLCN7* (confirmed by finding cosegregation among four other family members).

Patients in studies of multiple family members with the same phenotype leading to the discovery of new genes include two siblings with mitochondrial complex V deficiency, *ATP5A1*; one family with postaxial polydactyly, *ZNF141*; three siblings with early-onset multisystem neurodegeneration, *UCHL1*; and two members of a consanguineous family with macrocephaly and epiphyseal dysplasia, *KIF7*.

Studies of multiple unrelated patients with the same rare phenotype have led to the discovery of the following associations between phenotypes and genes: Myhre syndrome and *SMAD4*; study of four families of Tunisian origin with autosomal recessive cerebellar ataxia, discovered a new gene, *GBA2*, was the cause; and four unrelated people with trismuspseudocamptodactyly syndrome, *tropomyosin*.

Since no one publishes their failures or their nondiagnostic studies, examples of WES failures are hard to come by. Thus, how do you know if your patient would benefit from this approach? It is based on general principles. Features that may make WES more likely to be successful for patients include

- Multiple individuals in a family with the same phenotype
- DNA available on more than one affected
- DNA available on those clearly not affected
- Phenotype clearly definable
- Phenotype not thought of as multifactorial
- Phenotype might suggest candidate pathways
- Mendelian pattern clear
- Consanguinity

Features that reduce chances that WES will be successful in making a diagnosis include

- Only one affected individual available for testing within the family or across families
- Difficult to assign unaffected status to family members
- Phenotype undefined or poorly defined (i.e., immunologic workup not possible or all biochemical testing normal)
- Phenotype not usually attributable to single-gene process, for example, diabetes mellitus type II, hypertension, and obesity
- No clear Mendelian pattern in family and no consanguinity

There must be counseling for families and patients wishing to undergo WES.

WES requires DNA, usually from blood. There are issues concerning whose DNA would be needed—possibly that of estranged relatives or adopted-out siblings—that would make obtaining DNA difficult. Relatives may be unaware of their potential risks, and some may be reluctant to share personal medical work beyond the nuclear family. Other issues include costs: who bears those costs, and how long does the test take? Even in the most optimal of scenarios, a diagnosis is made only about 30% of the time. Why? There is no real data to provide numbers about the probability of getting an answer. WES does not read the whole genetic code; some types of mutations will be missed. There is the inability of WES to distinguish real mutations from nonmutations, and finally, there is lack of information about many genes' normal function.

What might one learn from WES? Possibly absolutely nothing. Even if one could get a clear diagnosis, that would not necessarily mean that better treatments would be available. One may ask how important is it to have a diagnosis. What about variants of uncertain clinical significance? There will be lots of those, and most either cannot be resolved or else will require filtering across multiple family members as the most useful tool for eliminating these. There are also "incidental findings," which are significant genetic mutations that have nothing to do with the diagnostic odyssey search but pop up nevertheless. With regard to incidental findings, the American College of Medical Genetics has advised that if mutations are discovered in one of 56 specific genes of people having WES, they should be told. These are genes that mostly indicates a predisposition to cancers and sudden cardiac arrhythmias. Controversy exists about implementation of this one-size-fits-all guidance. According to Noralane Lindor, MD, consultant, Department of Health Sciences Research, Mayo Clinic in Arizona, "Incidental findings need careful further discussion, family by family, lab by lab. (The Mayo Clinic) has developed a counseling process for having more nuanced conversations about this."

Thus, WES has a proven role in establishing diagnoses in rare disorders. Transparent individual evaluation of each case and family before launching WES, and efforts to support personal autonomy in treatment decision making are important, and evaluation must provide realistic estimates of the chances of successful diagnoses for those considering WES.

PERSONALIZED HEALTHCARE

According to David Ginsburg of the Life Sciences Institute at the University of Michigan, next-generation sequencing will be transformational, and WGS will replace conventional newborn sequencing. Additionally, genomic testing has revealed the diagnosis of a number of Mendelian or single-gene genetic disorders, including cystic fibrosis and muscular dystrophy. Once a familial mutation has been identified:

> Testing of nearly perfect sensitivity and specificity is available for at-risk family members, including prenatal and preimplantation diagnosis. These advances have led to the elimination of select autosomal recessive diseases in specific populations, such as β-thalassemia of the Mediterranean and Tay-Sachs disease among Ashkenazi Jews (Ginsburg, 2011, 17).

Unfortunately, with the exception of Gleevec for chronic myeloid leukemia (CML), treatment for these disorders for which genetic pathogenesis has been understood remains largely elusive.

This is where personalized healthcare (PHC) comes into the picture. According to Burnette et al., PHC has been proposed as a rational approach to fill the need for an innovative, cost-effective model for healthcare delivery:

> PHC is a coordinated, strategic approach to patient care that broadly applies the concepts of systems biology and personalized, predictive, preventive, and participatory care (known as prospective health care or P4 medicine), and uses available technologies to customize care across the health continuum from health promotion and prevention, to detection and treatment of disease. While PHC anticipates using the vastly improved predictive tools created as a consequence of genomic technologies,

it does not require them to get started. PHC is designed to proactively change the trajectory of disease development by using available predictive health risk assessment capabilities and planning while intensively engaging the patient in coordinated approaches to their care. Although personalized medicine is often equated with genomic medicine, PHC, as defined herein, takes a broader view and embraces current capabilities and tools to provide the best strategically planned predictive care. Nonetheless, the PHC approach provides a portal for the clinical adoption of genomic technologies as they are validated (Burnette et al., 2012, 233–234).

PHC generally involves evaluating patient health status, tracking disease development, and providing a therapeutic plan based on personalized predictive tools. As with personalized medicine in general, the objective is to go from reactive healthcare to proactive healthcare, which is befitting for a rapidly aging population who will ultimately obtain their healthcare services through Medicare. Furthermore, for those with preventable chronic disease, studies (see below) have shown that they would benefit from PHC. Thus, genomics and personalized medicine are providing new tools that allow for preventive and more precisely delivered healthcare.

As omic technologies are being developed, they will be incorporated into the personalized health plan. Using outcomes (e.g., biomarkers), drug metabolic markers, and treatment selection, the predictive tools of personalized medicine will meet patient needs. This has been shown in a study on the implementation of a PHC application, called My Personalized Risk EValuation, ENgagement, and Tracking Plan (My PREVENT™ Plan). According to Burnette et al. (2012, 235), "the primary aim of My PREVENT™ Plan is the identification of the patient's proximate and longer-term health risks and needs, chronic disease management, patient engagement, and coordination of care. The plan will embrace personalized therapeutic tools, where applicable." Providers noted that when implementing PHC, information for patients was particularly useful for determining

health status, social determinants of health, and patient preferences.

FINDING VARIANTS OF SIGNIFICANCE IN CLINICS AND SEQUENCING CENTERS

Advances in sequencing technologies allow for the provision of genome-scale data to oncologists and geneticists caring for pediatric cancer patients. The goal of the BASIC3 (Baylor Advancing Sequencing into Childhood Cancer Care) study is to determine the clinical impact of incorporating Clinical Laboratory Improvement Amendments (CLIA)-certified tumor and constitutional exome sequencing into the care of children with newly diagnosed solid tumors. The purpose of BASIC3 is to integrate information from CLIA germline and tumor exome sequencing. The following case studies illustrate how BASIC3 works.

In case study 1, a patient appeared at Baylor with bone tumor osteosarcoma with metastasis. Cytogenetics, which reveals a portrait of all the chromosomes, showed the chromosomes in a completely aberrant state. Treatment involved chemotherapy and tumor resection, and the prognosis was poor. Tumor exome results revealed a somatic $p53$ mutation, a somatic $TSC2$ mutation, and a two-base-pair deletion indicating a truncating mutation. However, the database did not reveal a $TSC2$ mutation in bone tumors. TSC2 stands for the tuberous sclerosis complex, which is a main inhibitor of mTOR in a signaling pathway in cells. There were mTOR inhibitors in the clinic, and clinicians at Baylor were considering using them for this patient. In terms of targeting mTOR in TSC-related tumors, there was Food and Drug Administration (FDA) approval for everolimus, also known as Afinitor. mTOR inhibitor therapy in conjunction with chemotherapy proved to be an effective treatment for the patient.

Case study 2 demonstrates a patient with a huge heterogeneous tumor in the liver revealed through MRI with metastasis. Doctors did not know what kind of liver tumor it was. Tumor exome results showed an $nRAS$ mutation, normally seen in melanoma. Chemotherapy and a hepatotomy were performed. After the fourth cycle of chemotherapy, there was progression of lung metastasis. In terms of targeting ras, $PI3K$ and raf are downstream effectors. The drug sorafenib targets *raf*, which was used in therapy, thus matching clinical and genetic information to find out what works.

Case study 3 involved a pediatric patient with no family history of cancer, but the exam showed a large abdominal tumor in the kidney. Observed to be Wilms' tumor, treatment involved chemotherapy and abdominal radiation. Germline exome results revealed a small deletion in all reads in the tumor and a mosaic mutation in *WT1*. This indicated that the tumor emerged not from only a germline mutation but also from an inherited somatic mutation. The germline-inactivating *WT1* mutation was associated with Wilms' tumor, and patients have an increased risk of renal failure unrelated to a contralateral Wilms' tumor.

Thus, BASIC3 serves as an assessment of feasibility and utility of exome sequencing. It is a pipeline for clinical sequencing of patient tumor and blood samples, and will provide insight into the preferences of oncologists and families for exome reporting, with actionable tumor and germline results identified. However, there are other methods to identify germline mutations, such as a targeted panel, RNA-seq, and arrays, along with WGS, in addition to clinical exome sequencing. Therefore, it is not yet clear what the most efficient combination will be. Other questions yet to be answered concern the clinical interpretation of exome results: Which mutations are pathogenic? Which is the best drug to target a mutation, and when and how is the drug to be used? What evidence of potential benefit is needed to use the drug? How should incidental noncancer findings from the germline be handled? And how confident do clinicians need to be to subject the patient to toxicity?

Once a target has been found, the issue of cost arises. Many targeted drugs cost $5000–$10,000 a month, and most are not covered by insurance. Additionally, many drugs are not yet FDA approved for any indication, and early-phase clinical trials open and close frequently. Considering that the cost of sequencing (exome sequencing currently costs $1500) and therapy is so expensive, are both sequencing and treatment worth it? Furthermore, what about heterogeneity with tumors expressing multiple mutations in different tumor cells? Gene panels are extremely important in this context, since they can provide actionable mutations, particularly for solid tumors. The starting sample is important, for tumor purity influences the output for detecting driver mutations that have clinical utility.

REFERENCES

Burnette, R., Simmons, L.A., and R. Snyderman. 2012. Personalized health care as a pathway for the adoption of genomic medicine. *Journal of Personalized Medicine* 2:232–240.

Ginsburg, D. 2011. Genetics and genomics to the clinic: A long road ahead. *Cell* 147:17–19.

CHAPTER 9

The regulatory landscape

We now have a world turned upside down.

Robert Califf, MD
Deputy commissioner, FDA

From Pfizer's Xalkori, targeted for non-small-lung cancer cells harboring the ALK mutation, to Genentech's Herceptin for HER2-positive cells in breast cancer, the Food and Drug Administration (FDA) has played a central role in the development of personalized drugs and companion diagnostic tests, ensuring their safety and efficacy. However, the FDA has faced many challenges in coordinating the approval of the drug–test combination and regulating the clinical trials for drugs. The FDA has also made changes to the enforcement discretion requirement for laboratory-developed tests (LDTs). Next-generation sequencing (NGS) has also challenged the regulatory landscape, leading to new regulations concerning the intended use of genetic variants. In all these areas, the FDA is utilizing new approaches to meet innovation in drug and medical device development while still accommodating patient needs for clinical effectiveness and safety.

Robert Califf discusses the new ecosystem and priorities for the FDA for precision medicine. Califf talks about how in the old ecosystem, academics worked on basic research, companies made drugs, insurers paid, and patients were passive recipients of brilliance. But with personalized medicine, patients are now involved more deeply as consumers using social media as advocates.

Califf notes that biomarkers are critical to drug development and personalized medicine. Getting from a qualified biomarker to qualified drug approval in tandem requires a lot of hard work. Califf sees the three big challenges in medicine as mobile information over time, such as wearable devices and glucose sensors; systems biology, where we can measure multiple markers in many dimensions; and issues of ethics and privacy.

The FDA has issued guidance statements for companion diagnostic test and *in vitro* diagnostic test approval; however, as personalized medicine continues to evolve and drug and diagnostic companies are constantly in the process of innovating during drug and diagnostic test development, the FDA has faced criticism and industry dissatisfaction for not keeping up with the advances in personalized medicine and technologies.

Companion diagnostic tests are regulated as medical devices, while targeted therapies such as Xalkori and Herceptin are regulated as drugs. Pharmaceutical companies collaborate with diagnostic companies to develop drugs in parallel with the diagnostic test. For example, in a report by *PGx Reporter* in February 2012, Turna Ray reported that Pfizer

[harmonized] the development and review of Xalkori and ALK mutation test. Pfizer had initially used laboratory-developed tests to gauge ALK mutations in patients, but then decided to work with Abbott Molecular to develop

111

a diagnostic kit that was submitted to the FDA.

To accomplish the simultaneous approval and market launch of Xalkori and Abbott's ALK mutation test, FDA's drug and diagnostics division had to work "very closely together" and hold "many, many meetings," Mansfield, former Commissioner of the FDA, said. "The diagnostic typically has a shorter review time than the drug, but sometimes the drug review is accelerated … so we had to come up with a process to allow that to happen" (Ray, 2012).

The FDA is also adapting to the changing nature of clinical trials brought about by the nature of personalized drugs. The FDA has made exceptions to its usual requirement for double-blind, randomized clinical trials by allowing for clinical trial participation for only biomarker-positive patients for targeted therapies, eliminating the requirement for participation of biomarker-negative patients in clinical trials. Many of these changes were allowed to bring about rapid approval of urgently needed personalized oncology medicines and tests.

The FDA has changed its infrastructure to address issues associated with personalized medicine. According to the FDA Group:

This has included restructuring the offices responsible for the approval of drugs, devices, and biologics and creating programs to accommodate research associated with personalized medicine. The Center for Biologics Evaluation and Research has launched initiatives to create a review process that accommodates genetics and other innovative technologies. Programs such as the Voluntary Exploratory Data Submission program enable pharmaceutical companies to discuss with the Food and Drug Administration issues associated with personalized medicine outside of the standard approval process. These changes define the responsibilities of different departments so that the reviews of personalized medical products can be timely and consistent (FDA Group, 2015).

LABORATORY-DEVELOPED TESTS

The vast majority of currently marketed genomic tests are laboratory-developed tests (LDTs) that have not undergone FDA review. These tests either are fully developed by a laboratory or use a purchased analyte-specific reagent (ASR), which the laboratory that carries out the test as a service configures into an assay. Such LDTs are not shipped for use outside the originating site.

These tests are considered medical devices under the Federal Food, Drug, and Cosmetic Act (and the laboratories are considered device manufacturers), but the FDA, considering the controls provided by the ASR regulations and the certifications for high-complexity laboratories under Clinical Laboratory Improvement Amendments (CLIA) (see below) to be sufficient, has used its enforcement discretion and not required marketing applications. In September 2006, the FDA issued guidance entitled "In Vitro Diagnostic Multivariate Index Assays," which explains that certain complex test systems, including certain genomic test systems, are class II or class III medical devices requiring 510(k) submissions.

FDA laboratory test development guidance has declared recent initiatives in the realm of LDTs, including

- Draft guidance released in October 2014; final guidance pending
- Risk-based oversight by the FDA
- Ending of enforcement discretion
- Phase-in over nine years
- Premarket notification and approval for high-risk LDTs
- Requirement for submission of data on both analytical and clinical validity
- Carve-outs for "traditional LDTs," rare diseases, and unmet needs
- Carve-in for any LDT used for a companion diagnostic indication
- Adverse event reporting for all LDTs
- Quality system requirements for medical devices (not CLIA)

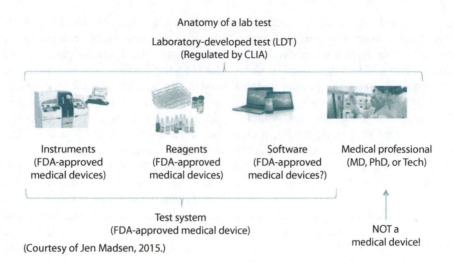

Anatomy of a lab test

Laboratory-developed test (LDT)
(Regulated by CLIA)

Instruments
(FDA-approved
medical devices)

Reagents
(FDA-approved
medical devices)

Software
(FDA-approved
medical devices?)

Medical professional
(MD, PhD, or Tech)

Test system
(FDA-approved medical device)

NOT a
medical device!

(Courtesy of Jen Madsen, 2015.)

- Grandfathering of currently marketed high-risk LDTs (Courtesy of Jen Madsen, 2015)

ANALYTE-SPECIFIC REAGENTS

In 1997, the FDA defined a group of reagents known as ASRs and classified them (with some exceptions that are not germane to genomics) as class I medical devices. This meant that ASRs were subject to "general controls," such as manufacturing and labeling requirements, but did not have to have applications cleared or approved by the FDA to be put on the market. The FDA accomplished this by publishing three regulations that (1) defined and classified ASRs (21 CFR 864.4020); (2) imposed restrictions on their sale, distribution, and use (21 CFR 809.30); and (3) established requirements for ASR labels (21 CFR 809.10[e]).

In the regulations, ASRs were defined as antibodies, both polyclonal and monoclonal, specific receptor proteins, ligands, nucleic acid sequences, and similar reagents that, through specific binding or chemical reactions with substances in a specimen, are intended for use in a diagnostic application for identification and quantification of an individual chemical substance or ligand in biological specimens (21 CFR 864.4020). Thus, an ASR is a building block for an assay,

but is not a test system. Subsequent to publication of these regulations, the FDA observed a broadening of the use of ASRs beyond the stated parameters. These distinctions were, and remain, important to the genomic testing community; they have elicited extensive comment. Undoubtedly, the boundary between ASRs and test systems will continue to be clarified (Woodcock, 2013).

CLINICAL LABORATORY IMPROVEMENT AMENDMENTS

In 1988, Congress passed the CLIA, to be administered by Centers for Medicare & Medicaid Services (CMS) (https://www.cms.gov/CLIA/). This law established quality standards for all laboratory testing, to ensure the accuracy, reliability, and timeliness of patient test results regardless of where the test is performed. The ASR regulations just discussed stipulated that the only clinical laboratories to which ASRs could be sold were those qualified under CLIA to perform high-complexity testing (or, alternatively, are regulated under the Veteran's Health Administration Directive 1106) (21 CFR 809.30[a][2]).

Currently, most genetic testing is performed by such laboratories. The oversight under CLIA relates to the quality of performance of laboratory testing and does not extend to the evaluation of clinical utility (i.e., the magnitude of benefits vs risks of performing the test in a specific population) of a given assay.

Although analytical and clinical validity of such tests are of foundational importance, much of the controversy over regulation of these tests will center on how much demonstration of clinical utility is required, because generating such evidence is often very expensive and time-consuming. There is also ongoing controversy about the degree of oversight of genetic testing under CLIA (Hudson et al., 2006), in particular, whether there should be a specific genetic testing specialty area that would incorporate proficiency testing for genetic tests.

To summarize the current federal regulation of genomic testing, most genetic tests are on the market as LDTs. Relevant laboratories are subject to quality standards under CLIA. This has come about as a result of the ASR regulations, coupled with the FDA's use of enforcement discretion for LDTs. Much of the controversy over the use of genetic tests in healthcare stems from the fact that the FDA has not actively regulated such tests, and CLIA does not call for clinical utility. The FDA has approved a number of genetic tests, including tests for drug-metabolizing enzyme polymorphisms, generally based on information available in the scientific literature or data from retrospective testing of stored samples. The FDA stated its intent to change its policy on enforcement discretion (Woodcock, 2013).

CHALLENGES OF MEDICAL DEVICE FRAMEWORK FOR NGS

The FDA regulates NGS as a medical device, including instruments, reagents, and clinical variant data. FDA used the *de novo* process to review the Illumina system in 2013 and the 23andMe Bloom syndrome test. While NGS challenges the "intended use" framework, the FDA has stated that NGS is "broad and indication-blind testing" that (1) can identify an unlimited number of mutations however unable to assess analytical performance on all mutations and (2) can identify very rare mutations, declaring that it is not feasible to run clinical trials on each one.

The FDA is planning a new approach: "new regulatory approaches will be needed to enable the Agency to provide appropriate oversight, in a way that is more suitable to the complexity and data-richness of this new technology," according

to the Jen Madsen of Arnold & Porter LLC. To ensure their analytical performance, sequencing technologies must rely on quality-based standards for NGS test performance that would be created in collaboration with experts in genomics and used to ensure test accuracy and reliability. To ensure the clinical validity of NGS, technologies would use high-quality curated genetic databases that provide information on genetic variants and their association with disease, for example, ClinVar and ClinGen. A precedent existed in the 23andMe test for Bloom syndrome where, in February 2015, the FDA used current regulatory authorities to down-classify the test to class II to approve carrier screening tests through the 510(K) process, signaling flexibility.

FDA PHARMACOGENOMICS DRUG LABELING

Lawrence Lesko, PhD, at the Center for Pharmacometrics and Systems Pharmacology at the University of Florida in Orlando, states that for more than a decade, the FDA has been urging adoptions of pharmacogenomics strategies and pursuit of target therapies and harmonizing field labels as an important part of FDA deliberations. Since the labeling of 6-mercaptopurine with thiopurine S-methyltransferase (TPMT) in 2002, genomic information to the FDA is provided in order to make the FDA more genomic friendly. In 2005, a guidance on voluntary submissions was issued by the FDA, while 100 submissions to the FDA were made, making the Agency an enabler of personalized medicine.

Labels form the legal basis for approving and marketing drugs, and the critical role of drug labels is as follows:

- They constitute the legal basis for prescribing and marketing medicines, but not dictating the practice of medicine
- Their primary purpose is to give healthcare professionals the information they need to prescribe appropriately pharmacists
- Sponsors who use labels propose language as their summaries of new drug development and negotiate with the FDA early-phase drug development program, biomarkers, final label negotiated label

- Labels also constitute the most common way to warn about safety risks and inform tertiary drug information sources commercial databases
- They also are prerequisite to third-party reimbursement

Labels are tightly regulated. According to 21 CFR 201.57, "if evidence is available to support the safety and effectiveness of the drug only in selected subgroups of the larger population with a disease, the labeling shall describe the evidence and identify specific tests needed for selection and monitoring of patients who need the drug."

Specific warnings that have been adopted by the FDA over the decade include those for eliglustate for Gaucher's disease; CYP2D6 tests, which test for ultrarapid metabolizers that do not receive the drug; trastuzumab and Her2, which constituted the first codevelopment of a biomarker and drug; thioridazine and CYP2D6, which indicated use of the first warning; 6-MP TMPT, the first dose stratification by genotype; erlotinib and EGFR; panitumamab and KRAS; warfarin and CYP2C9, the first detailed genotype-guided dosing; vemurafenib and BRAF, a specific FDA approval test with an FDA label, the prototype for FDA breakthrough therapy; and ivacaftor, the first breakthrough therapy for 5% of cystic fibrosis patients who were positive for G551D mutations, approved without a companion diagnostic.

According to Lesko, there are 141 drugs with 45 genomic biomarkers, and while there are no national standards for labels, 35% of labels have informational value, while 40% are actionable with no test and 21% are actionable with a test. There are two ways a label can be instituted, with codevelopment of a drug and test that facilitates the linking of the label with the drug, and the development of the test after the drug reaches market, which makes labeling more difficult.

REFERENCES

FDA Group. 2015. Personalized medicine and the FDA. Westborough, MA: FDA Group, September 15. Available from http://www.thefdagroup.com/thefdgroup-blog/2015/09/personalized-medicine-and-the-fda/.

Hudson, K.L. 2006. Genetic testing oversight. *Science* 313(5795):1853.

Ray, T. 2012. FDA official says personalized medicine regulation taking shape following Zelboraf, Xalkori success. *PGx Reporter.* Available at https://www.genomeweb.com/mdx/fda-official-says-personalized-medicine-regulation-taking-shape-following-zelbor.

Woodcock, J. 2013. Regulatory issues for genomic technologies. In *Genomic and Personalized Medicine*, Ginsburg, G., and Willard, H., eds., 422–432. 2nd ed. Waltham, MA: Academic Press.

CHAPTER 10

Translational personalized medicine: Molecular profiling, druggable targets, and clinical genomic medicine

One of the many concerns about implementing personalized medicine stems from the question, is there enough evidence on the clinical utility of personalized medicine to fulfill its expectations? A number of projects under way are designed to determine the clinical efficacy of personalized medicine. The Coriell Institute for Medical Research, Camden, New Jersey, currently presides over the Coriell Personalized Medicine Collaborative (CPMC). This multipronged study consists of the genetic testing of patient DNA samples, interviewing patients (completing health questionnaires), genetic counseling, and assessing patient disease risk. Involving more than 9000 patient participants and a research team of geneticists, bioinformaticians, and genetic counselors, the CPMC aims to deliver on the idealized clinical benefits of personalized medicine. With an informative website on diseases for genetic tests that can reveal disease risk, and for disease-treating drugs that are on the pharmacogenomics spectrum, such as thiopurines and warfarin, the collaborative seeks to publicize the links between genomic testing and clinical efficacy. The website also addresses genetic privacy and the security measures of the study. Michael Christman, PhD, president and CEO of the Coriell Institute, calls this "personalized medicine in practice." The Coriell's Genome Center is high throughput and CLIA certified. Coriell's pharmacogenetics

evidence evaluation processes translation pharmacogenomics discoveries into the clinic. Christman cites a Medco study where 98% of physicians agree that patient genetic profiles may influence drug therapy, with only 13% of physicians having ordered a pharmacogenetics test in the previous six months and only 10% who believe they are adequately informed about pharmacogenetics testing.

The CPMC recently received a $10 million Air Force award to enroll 4500 participants and aggregate electronic medical record data to study obesity, cardiopulmonary fitness, and sleep medications. The study found that the Air Force gets significantly less sleep ($n = 1262$) and has a larger percentage of individuals taking sleep medication. The only finding was that sleep deprivation leads to several health issues, including increased risk for posttraumatic stress disorder.

The University of Chicago's Center for Personalized Therapeutics draws on a similar study in its 1200 Patients Project. With the caption "Can a simple blood test lead to better medical care?" the 1200 Patients Project demonstrates how genetic testing can be incorporated into the clinical decision-making process (e.g., prescribing medications) as a way to individualize therapies. Their research "could help develop a new medical system model for personalized medical care," (1200 Patients Project website) and stands at the forefront of implementing pharmacogenomics testing in conjunction with prescribing drugs to prevent adverse reactions and predict accurate dosing. This, again, will advance personalized medicine's association with evidence-based medicine. If insurance companies are inclined to pay for evidence-based treatment, they may be more likely to pay for genomic testing and medicine.

The University of Arizona (UA) Health Sciences seeks to establish a healthcare delivery platform with clinical partners that individualizes healthcare by incorporating "omics" approaches into clinical diagnosis and treatment, patient stratification, and clinical management, with improvements in individual and population health, while reducing healthcare costs.

UA precision health strategies (Table 10.1) include genome-based evaluations for increased diagnostic precision, targeted therapies that improve pharmacological efficacy and decrease toxicity, and optimized clinical management based on omics stratification and molecular diagnostics. UA genomics technology and innovation include

- Prosigna Breast Cancer Prognostic Gene Signature Assay, a gene expression–based test that analyzes the activity of 50 genes in early-stage hormone receptor–positive breast cancer. Nanostring software is used to generate custom clinical reports.
- Fragile X testing offered by Asurage (Food and Drug Administration [FDA] approved). Analysis and patient reporting are handled through the GeneMarker software suite.
- TruSite solid tumor testing, which uses next-generation sequencing to test selected regions of 26 genes frequently mutated in cancer that are actionable and billable.

Novel UA Health Sciences approaches in genomic medicine are the rendition of the complex network of relationships between disease and the human genome. Many disease-associated genetic variants fall outside of protein coding genes, instead affecting the genome's regulatory circuit by modifying DNA switches that control gene activity. A major area of interest is the characterization of repetitive, long interspersed mobile elements (LINEs) as mediators of disease and biomarkers of pulmonary disease.

The Mount Sinai School of Medicine's Charles Bronfman Institute for Personalized Medicine is also making heavy inroads in providing evidence for the clinical effectiveness of personalized medicine and genetic testing through two studies: BioMe and CLIPMERGE. Erwin Bottinger, MD, director of the Institute, is a clinician–researcher who came to Mount Sinai in 2004 and presented the idea to start an institutional effort in personalized medicine.

At the time it was very forward thinking because people hadn't dared to talk about it, so we were way ahead in discussions. That led to [the generation of] a philanthropic gift which allowed us to launch the Charles Bronfman Institute for Personalized Medicine. The objectives of the Institute are threefold, first to make an unprecedented resource for patients that are receiving clinical care at Mount Sinai, and consent as part of a research project that they will give a tube of blood from which we will extract DNA and that DNA can then be sequenced and genotyped. The data from the sample

Table 10.1 UA precision healthcare pipeline. From omics to product development to healthcare delivery

Omics innovation	Product development	Regulatory science and reimbursement	Healthcare delivery
Genomics	Pharma/Biotech	Curricula	Patients
Proteomics	Medical Devices	Regulators	Banner University Medical Center
Metabolomics	Molecular Diagnostics	Payers	Physicians
Bioinformatics		Health IT	Nurses
			Genetic Counselors

Source: Courtesy of Ken Ramos, 2015. Arrowhead Personalized and Precision Medicine Conference.

can be linked with the electronic health record [EHR] going back into the past and going forward into the future as they come back to see the doctors at Mount Sinai; every encounter is creating a new record. With this launched, we created a biobank which became BioMe in September 2007. We have a dedicated group of clinical research coordinators whose only job is enrolling patients. Fast-forward, and we now have enrolled 30,000 Mount Sinai patients. And with that, these patients come to see doctors at Mount Sinai with common ailments such as diabetes, high blood pressure, asthma, and many other common conditions. Among those 30,000 patients, we have 14,000 with high blood pressure; we have 8000 with diabetes, 4000 with asthma, and large numbers of cases with these individual diagnoses. That allows us to address the genetics of common diseases by applying GWAS [genome-wide association studies] approaches. We have now made contributions from the BioMe biobank dataset to over 40 research studies, which are now arriving in publications. In addition, on the basis of biobank availability, we have recruited top-level faculty investigators who are using the biobank resource for research on population genetics, for instance, trying to turn genetic heterogeneity into a strength in biodiscovery. This had led to US\$50 million in research funding from the National Institutes of Health (NIH) directly linked to the biobank. We continue to generate into new insight, knowledge and approaches.

CLIPMERGE [another study at the institute] is the flip side of the coin, and concerns how we can get information back into clinical care workflow. CLIPMERGE stands for Clinical Implementation of Personalized Medicine from Records and Genomics. It is essentially a software that provides IT solutions that can interact with electronic health records. What does that mean? While the doctor is with a patient and has the EHR open, it allows relevant information from the EHR to flow to the CLIPMERGE system. With particular genotype variant data from the patient available, and when a new order for medication is provided, CLIPMERGE can give guidance through genetic information as to whether that is the right order. If we have the genotype information, then the clinical decision support system advises the doctor that this order is problematic for the patient: the medication doesn't work or causes severe side effects for the patient. It works when implemented [as many studies have attested to].

Vanderbilt University School of Medicine's personalized medicine initiative, under the direction of Dan Roden, MD, is involved in two projects designed to advance the clinical utility of genomic medicine. Roden states that there are two approaches in the pharmacogenetics world to getting data in the patients' records that will allow it to be used at some future date. "One is reactive genotyping: somebody has a drug like clopidogrel prescribed; the system says you should do a test and the system will tell you how to take care of that particular result. Another way is to look forward to the day when everybody's genetic make-up is embedded in their electronic medical record. This is preemptive genotyping: the physician obtains the genotyping done before it is needed. It sits in the chart, is called upon when needed, and the physician then prescribes the appropriate dosage of drug."

Vanderbilt's efforts in this area manifest in BioVU and PREDICT. BioVU is a biobank, and it is the biggest project that Vanderbilt is involved in. BioVU takes DNA samples that are obtained in the course of ordinary patient care, extracts that DNA, and couples it to an electronic medical record. The results are patients with clinical information scrubbed by identifiers attached to a DNA sample. So, the DNA sample is attached to a "deidentified" person. That biobank currently has more than 185,000 samples from individual subjects. According to Roden, it is the largest biobank in the world at a single academic medical center. The availability of BioVU has allowed Vanderbilt to become part of multiple research networks, including the NIH, to do discovery and implement genetic and pharmacogenetics variants in patients' electronic medical records.

On the identified side, Vanderbilt has a project called PREDICT (Pharmacogenomic Resource for Enhanced Decisions in Care and Treatment) that takes the first steps toward this vision of preemptive pharmacogenomics testing. Clinicians identify patients at high risk for abnormal responses to drugs like clopidogrel or warfarin, which is then put into the patients' charts. "When the drug is prescribed, the information is there. PREDICT has already involved over 15,000 patients whose identity we know."

BASIC RESEARCH

With new advances being made on a daily basis in genomic sequencing, genetic studies, and basic molecular research, the twenty-first century may go down as the age of genomic research. Basic genomics research has been conducted on many levels to implement personalized, preventive medicine: finding druggable targets in the genome through pharmacogenetics research, finding genetic polymorphisms through GWAS and the haplotype map project (haploid genome); the discovery of clinical biomarkers for disease and illness; and the discovery of somatic mutations harbored by tumors. Each of these avenues of research has advanced personalized medicine to a degree, reflecting different scientific methodologies, techniques, and rationales. At the Harvard Medical School, Mayo Clinic, Washington University School of Medicine, and other institutions, basic research in personalized medicine has been advancing at an astounding pace.

Druggable genome

"The Druggable Genome Is Now Googleable" read a November 2013 headline of Bio-IT World, a website devoted to next-generation technologies and personalized medicine. The article delves into a searchable database that would obviate the need for a bioinformatician by the pharmacogenomics researcher. Obi and Malachi Griffith developed the idea of the Drug Gene Interaction Database (DGIdb), a free, searchable online database of drug–gene associations. The Griffith brothers claim that this database can be used by the non-informatics expert. Entering search terms brings up a chart of drug–gene interactions that are culled together from public databases such

as DrugBank, the Therapeutic Target Database, and PharmGKB. It was a labor-intensive activity to set up the DGIdb, but the result is a Google-like website that can reveal drug–gene interactions through search filters (Krol, 2013).

As the article also explains, pharmacogenomics research has quickly been expanding to reveal drug–gene targets for new and existing drugs. According to McKinnon et al. (2007, 751):

Over recent decades, basic research has yielded a large volume of data on many potentially clinically relevant genetic determinants of drug efficacy and toxicity. Until recently, most examples involved genes encoding drug-metabolizing enzymes, particularly the cytochromes P450. More recently, rapid advances in genomic technologies have enabled broader, genome-wide searches for determinants of drug response. In parallel with these pharmacogenetic studies, a new drug discovery platform, termed pharmacogenomics, has emerged which utilizes genetic information to guide the selection of new drugs most likely to survive increasingly demanding safety and efficacy assessments. Together, these advances are widely promoted as the basis of a new era of drug-based therapeutics tailored to the individual.

According to Walko and McLeod (2009), dosing for drugs, especially in oncology, is performed on the basis of patient body surface area calculations, which can lead to erroneous results since body surface area corresponds poorly to drug pharmacokinetic activity. In a search for better methods of developing drug dosage, certain drug–gene pairs have been discovered that lead to different metabolic activities of drugs in the body. This individualized therapy approach is a better predictor of drug response, and would prevent adverse drug reactions and toxicity. Consider the following examples of drugs whose dosages resulting from pharmacogenetics research have been altered to increase drug efficacy and safety.

Mercaptopurine, thioguanine, and azathiopurine are thiopurine antimetabolites that inhibit DNA synthesis by incorporation of incorrect bases (thioguanine nucleotides). These immunosuppressive agents are used in the treatment of leukemias and autoimmune disorders. The drugs are inactivated by the enzyme thiopurine S-methyltransferase (TPMT). Three alleles in the *TPMT* gene encoding this enzyme account for 95% of the mutant alleles seen in the Caucasian, Asian, and African–American populations (Walko and McLeod, 2009, 153).

These three alleles result in the fast degradation of TPMT, which causes the enzyme to be deficient.

Approximately 10% of Caucasians and African–Americans are heterozygous for one of these alleles, which results in intermediate TPMT activity, and 0.3% are homozygous and nearly completely deficient in TPMT. When patients who are homozygous for one of these alleles are administered standard doses of thiopurines, all develop severe hematologic toxicity.

- *Irinotecan*: The mutant *UGT1A1* allele is associated with increased neutropenia and diarrhea in patients receiving irinotecan, and the FDA has recommended dose reduction in patients who are homozygous for this allele.
- *Tamoxifen*: Patients receiving tamoxifen who are classified as CYP2D6 poor metabolizers, on the basis of either genotype or concurrent drug therapy, are likely to have recurrence and lower disease-free survival than extensive metabolizers or patients who are homozygous for the wild-type *CYP2D6* allele.
- *Warfarin*: Warfarin dosing can be influenced by a patient's height, age, and both *VKORC1* and *CYP2C9* genotypes, which can account for 54.2% of the variability in warfarin dose requirements and could potentially improve the quality of care provided for patients receiving this anticoagulant. However, studies on the utility of using genes to predict what dose to use for warfarin have not come to a firm conclusion. (Walko and McLeod, 2009, 153)

GWAS AND THE INTERNATIONAL HAPMAP PROJECT

According to Baranov (2009, 71):

The key role in the study of genetic polymorphisms is played by the international project for the study of the haploid human genome—Haploid Map (HapMap). The purpose of the project is to obtain a genetic map of the distribution of single-nucleotide substitutions (SNP) in the haploid set of all 23 human chromosomes. The main thrust of the project consists in that the analysis of the distribution of known SNP in individuals for several generations revealed that SNPs that are adjacent or closely located in the DNA of one chromosome are inherited as SNP blocks. This block represents a SNP haplotype—a set of alleles located on the same chromosome (hence the origin of the name of the project HapMap). Thus, each of the mapped SNP acts as an independent molecular marker. By linking these SNP markers with the manifestations being investigated (disease symptoms), the most likely location of candidate genes' mutations (polymorphisms) associated with a particular multifactorial disease are identified. Typically, there are 5 or 6 SNPs closely linked with the already known Mendelian traits selected for mapping. Well described SNPs with a frequency of rare alleles of not less than 5% are called marker SNPs (tagSNP). It is assumed that, ultimately, in the process of the project's implementation only about 500,000 tagSNP of the approximately 10 million SNPs present in the genome of every individual will be selected. But this number is more than enough in order to map with SNPs the full human genome and identify new candidate genes linked with various multifactorial diseases.

As discussed in Chapter 2, positional cloning, with its use of pedigrees and markers, served as a means to find linkages between chromosomal regions and diseases for multigene diseases. But, this turned

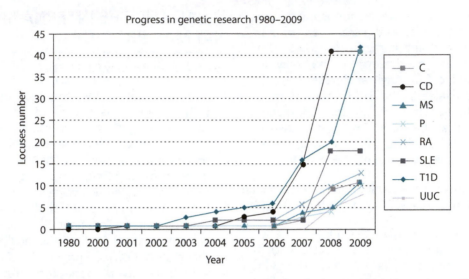

Figure 10.1 Progress in searching for genes of multifactorial disease of the immune system. C, celiac disease; CD, Crohn's disease; MS, multiple sclerosis; P, psoriasis; RA, rheumatoid arthritis; SLE, systemic lupus erythematosus; T1D, type 1 diabetes; UUC, unspecific ulcerative colitis. (Reprinted with permission from Baranov, V.S., *Acta Naturae*, 3, 70–80, 2009.)

out to be a very cumbersome process that was soon replaced by the more advanced method of GWAS. GWAS are based on the technologies of the HapMap program and high-resolution biochips. According to Baranov (2009, 73):

> As a result of the HapMap Project, the distribution of thousands of polymorphic sites—single-nucleotide substitutions (SNP) was revealed in the human genome and haplotype maps were created— maps of stable combinations of SNP variation within a single stranded (haploid) DNA sequence. Another important technical achievement was the hybridization of high-density DNA biochips that enabled simultaneous genotyping of thousands of SNP sites in one DNA sample. With knowledge of the exact position of each SNP on the physical map of the haploid genome, it has become not only possible to identify a candidate gene, but also to identify all SNPs associated with multifactorial diseases. The basis of the GWAS method is the scanning of hundreds of thousands of markers located on all human chromosomes. Thanks to the haplotypes maps obtained in the HapMap project, modern chip design includes the maximum number of key SNP (tag SNPs) and permits to estimate the frequency of both individual markers and haplotypes for the entire length of the DNA molecule.

GWAS are confirmed through patient samples and replicative cohorts to prevent false positives and allow for low *p*-values. Data from GWAS for diseases is available on the NIH website. Data obtained for genes associated with certain diseases of the immune system is presented in Figure 10.1. GWAS have advanced personalized medicine through their discovery of single-nucleotide polymorphisms (SNPs) associated with certain common diseases, such as diabetes, and allowed for clinical biomarker discovery, the next topic.

CLINICAL BIOMARKER DISCOVERY AND SOMATIC MUTATIONS

GWAS paved the way for personalized medicine to detect variants associated with diseases. The next step was to find clinical biomarkers associated with diseases that could be detected in blood, tissues, and even through imaging. The following research studies indicate biomarkers discovered for certain diseases that could establish biomarker-positive and biomarker-negative distinctions among patient groups in clinical trials, better predict response to treatment, and oversee disease progression.

Alzheimer's disease

According to Kumar et al. (2013, 1):

A minimally invasive diagnostic assay for early detection of Alzheimer's disease (AD) is required to select optimal patient groups in clinical trials, monitor disease progression and response to treatment, and to better plan patient clinical care. Blood is an attractive source for biomarkers due to minimal discomfort to the patient, encouraging greater compliance in clinical trials and frequent testing.

Kumar et al. (2013) have discovered a class of molecules, small noncoding RNA molecules termed miRNAs, that may regulate genes through post-transcriptional gene silencing that can be used to distinguish between Alzheimer's disease patients and normal controls. They found that through the detection of miRNAs in the blood, AD patients could be differentiated from normal controls with greater than 95% accuracy (Kumar et al., 2013).

Atherosclerosis and cardiovascular risk

According to Colley et al. (2011, 27):

A large body of basic scientific and clinical research supports the conclusion that inflammation plays a significant role in atherogenesis along the entire continuum of its progression. Inflammation adversely impacts intravascular lipid handling and metabolism, resulting in the development of macrophage foam cell, fatty streak, and atheromatous plaque formation. Given the enormous human and economic cost of myocardial infarction, ischemic stroke, peripheral arterial disease and amputation, and premature death and disability, considerable effort is being committed to refining our ability to correctly identify patients at heightened risk for atherosclerotic vascular disease and acute cardiovascular events so that they can be treated earlier and more aggressively.

Colley et al. (2011) have identified serum markers of inflammation that lead to inflammation and atherosclerosis. They discovered that lipoprotein-associated phospholipase A2 (Lp-PLA2) emerged as a marker of cardiovascular risk, supported by epidemiologic studies. They also report the development of a drug therapy, darapladib, that inhibits Lp-PLA2, which might reduce the progression of coronary artery plaques. They conclude, "The growing body of evidence points to an important role and utility for Lp-PLA2 testing in preventive and personalized clinical medicine" (Colley et al., 2011, 27).

Colorectal cancer

Biomarker discovery could also be crucial in designing effective personalized, preventive care for colorectal cancer. Raymond DuBois, MD, PhD, executive director of the Biodesign Institute at Arizona State University, notes that "using biomarkers wisely to save healthcare dollars [in terms of] avoiding giving EGFR inhibitor because [patients] have a *ras* mutation would amount to a $400 million savings."

Basic molecular profiling has emerged as a path to personalized medicine. Elaine Mardis, PhD, codirector of the Genome Institute at the Washington University School of Medicine, a pioneer in the field, notes:

We have been going after mutations in cancer for a long time especially in [the] context of therapy. [Through] PCR [polymerase chain reaction] amplification and sequencing, the *EGFR* mutation was the first demonstration of a correlation between a gene and a therapy. We saw mutations in tyrosine kinase inhibitor section of gene that responded to drug [therapy].

Getting mutations of the cancer genome through next-generation sequencing has moved rapidly. At [the] end of 2008 was the first recorded mutation in [the] tumor patient through whole genome sequencing. Ten somatic mutations in [a] patient's genome [were found]. In acute myelogenous leukemia, the average number of somatic mutations is 12, so 10 is close.

At Washington University School of Medicine, Mardis and her team have

> Sequenced well over 2000 tumors, adult and pediatric, which is driving discovery. Every cancer genome is different. We are committed to every evaluation of a cancer patient genome because every patient is going to be different and individualized.
>
> We are using next-generation sequencing to aid in diagnosis: acute promyelocytic leukemia. This particular subtype is characterized by a particular translocation between chromosome 15 and 17 that always has a 93%–95% cure rate with a regimen of chemotherapy.

Mardis announced the launch of the genomics tumor board at the Washington University School of Medicine, where "we are hearing cases for cancer patients consented for genome sequencing. All sequencing takes place in a CLIA lab with pathology signing off. Education, decision making, and patient monitoring are [all involved]. We use that information to return to the oncologist, who may use to return [it] to patient care."

Other somatic mutations have been found in illnesses such as non-small-cell lung cancer (*EGFR*, a tyrosine kinase receptor gene, and *KRAS*, a guanosine triphosphatase gene) and gastric cancer (*CDH1*, germline E-cadherin mutation).

TRANSLATIONAL MEDICINE AND TECHNOLOGY

One area in which biomarker discovery has been accelerated at certain centers is personalized medicine. Centers for personalized medicine are emerging in major medical centers throughout the country and translating the basic science research of personalized medicine into bedside treatment. The Center for Individualized Medicine at the Mayo Clinic, formerly directed by Gianrico Farrugia, MD, discovers and integrates the latest in genomic, molecular, and clinical sciences into personalized care for each Mayo Clinic patient. It consists of patient care that results from whole genome sequencing. Farrugia asks, "Why offer an IM [individualized medicine] clinic? We can bring people into [a] separate place where they can communicate;

there are also well-defined patient selection criteria. The IM clinic provides genomic counseling; financial counseling; extraction and processing of DNA; bioinformatics strategies; expert review boards to analyze patient data, interpret results, and search for therapy options; a genomic tumor board; and a genomic odyssey board. Results and possible treatment options [are given] to each patient, where one report goes into medical records." The IM Center is also making inroads into pharmacogenomics testing and epigenomics.

Farrugia asks, "How many biomarkers have been approved by the FDA over the past 20 years? Less than 30. The pathways for getting biomarker discovery into FDA approval are quite tough." However, Farrugia notes that with the advent of next-generation sequencing, biomarker discovery has been accelerated.

The Partners Center for Personalized Genetic Medicine, in collaboration with Harvard Medical School, is located in Cambridge, Massachusetts. The Partners Center for Personalized Genetic Medicine has several components. One is the Laboratory for Molecular Medicine, which Heidi Rehm, PhD, directs. Rehm states:

> In that lab, we develop new genetic tests, either based on new information on the causes of diseases, or bringing new technologies into use to make tests more effective and efficient. A lot of our work is centered around development of diagnostic tools, understanding genetic causes of disease, hereditary as well as somatic cancer, and then we also have a core facility which supports researchers, genetic studies, and have a biorepository that involves collecting patient samples for discovery purposes. So that is lot of what the center is focused on.
>
> In terms of my own research and collaborative studies, we are developing genomic resources to support integrated genetics into medical practice, that includes a large initiative called ClinGen, Clinical Genome Resource Program, [through which] we are collaborating with three other groups to develop genomic resources to better understand [the] role of genomic variation in human health and disease. We

are doing that by centralizing knowledge: all the different laboratories that interpret variants and genome, and genetic tests, we are all gathering together [this information] to share that data into a common place to make it much easier to understand those variants, by sharing data. We are sharing both curated knowledge, so if I study a particular variant and do testing on it, gather all knowledge associated with it and classify it with respect to disease, if I share that with my colleagues it will make it easier for them to know what to do when they see that variant, so, if thousands and thousands of labs do that, it will really help us be more efficient and consistent about interpreting variants. If we also share patients' genetic sequencing, and their phenotypes, we can use that [to] better understand [the] role of variants in disease, as well as make new discoveries about genetic causes of disease. We annotate variants of [known and unknown significance]; by putting that into the public domain it allows researchers to study variants that are understood as well as variants that are not understood, and be able to compare ones that are understood to ones that are not understood, to develop tools to help predict [the] impact of novel genetic variants.

However, Rehm notes:

Most PCPs [primary care physicians] are not ordering genetic tests today, so they are not [the] ones using this information. But a lot of specialists are already using the centralized resource which we have been involved in creating. We are working collaborative with Robert Green, one of the faculty in [the] center on a pilot clinical trial called the MedSeq study. This is a randomized pilot clinical trial to evaluate the role of whole genome sequencing in clinical care. So patients are randomized to receiving whole genome sequencing or a standard family history, and then we

sequence those that got randomized to the sequencing arm, to interpret their genomes and return them to them, and then record their disclosures and do surveys to understand how they use that information, what kind of impact it had on their care, and any anxiety related to that. We are halfway through it. My lab is providing all the whole genome sequencing reports, putting all of that data for the patient. Another study is called BabySeq, really looking at [the] introduction of whole genome sequencing into [the] care of newborns. Healthy newborns as well as babies that are not healthy, that are in the NICU.

According to Robert C. Green, MD, MPH, Associate Professor of Medicine of Brigham and Women's Hospital and Harvard Medical School and principal investigator on the MedSeq Project, early results show that genetic variants for pathogenic diseases are mainly inherited autosomal dominant. They are documented in a one-page report alerting the physician of these variants and the category available in the EHR. Primary care physicians have ordered several tests, including EKG (*KCNQ1*), iron and ferritin studies (*HFE*), and ultrasounds (cardiovascular risk alleles for coronary heart disease), for WGS diagnosis of disease variants. According to Green, despite perceptions to the contrary, physicians are as prepared, or unprepared, for genomic medicine as they are for other medical innovations.

Table 10.2 lists the six-month healthcare utilization and costs for the primary care physician cohort.

Patients reported feeling "happy," "empowered," and "relieved" upon discovering their results from WGS.

Don Rule, CEO and founder of Translational Software, a molecular diagnostics company that aims to put personalized medicine into practice, based in Mercer Island, Washington, would like for physicians to receive the latest in pharmacogenetics information. Rule states:

A lot of excitement about genetics is that common diseases would be caused by common variants. We would just find those variants and we would be all set. This obviously hasn't been true. You

Table 10.2 Preliminary data on healthcare utilization and costs for primary care physicians in the MedSeq Project. FH, family history; GS, genomic sequencing

	6-Month healthcare utilization and costs—PCP cohort				
	FH (n = 32)		FH + GS (n = 32)		
	Total	Per patient	Total	Per patient	p
Laboratory Tests	74	2.31	87	2.72	0.82
Imaging Tests	30	0.94	22	0.69	0.75
Cardiology Tests	5	0.16	9	0.28	0.31
PCP Visits	18	0.56	21	0.66	0.63
Non-PCP Visits	61	1.91	73	2.28	0.51
Total cost		**$825**		**$1161**	**0.49**

Source: Courtesy of Robert Green, 2015.

look at companies such as Navigenics and 23andMe; they can find correlations but not causations. We had the insight that in adverse drug reactions there is a relation between common disease and common variants. Houda Hachad [PharmD, scientific due diligence officer of Translational Software] has put together a panel of about 60 variations that have pretty well-known impacts on prescriptions. And what we do is integrate with the lab and get the genetic test results from them and based on those test results we create reports that show them what the implications are for the patient.

Rule discusses how useful Translational Software's services are to physicians. They look at themselves as sort of the supply chain that integrates the knowledge base of genetics with the genetic tests results and makes those useful to the doctor. They are looking to position themselves in such a way that a doctor of ordinary experience can use what they do. What that means is providing things in plain language; often what one sees from the lab, they publish as a genetic test result, and then provide generic information about what it means. What Rule's company tries to do is tailor directly to the patient and say, for a person of this genotype, what is the guidance the physician needs to know on how to dose him or her?

Rule notes that this information must be conveyed to the doctor in qualified terms. Some of the things Rule and Hachad are learning is that doctors do not like it if one is too in the imperative; in other words, one cannot say, "Avoid this medicine," but rather, "Consider finding an alternative to this medicine." Rule observes that his company needs to be careful that they are not diagnosing, but providing the best guidance that is available. They have had very good reaction from doctors at this point; however, there is a bit of tension between the labs and doctors in that the labs would love to tell you everything they know about molecular biology, because they are proud of what they do and find it fascinating. Rule finds that doctors are really busy and feel they need to see 40 patients a day to maintain their standard of living. They want a lot less; they want very concise, succinct information about considerations of patients. That's where Translational Software goes through continual evolution of reporting so they can figure out what is minimum and can say what is useful and actionable by the doctor. "We really focus on the implications of the genetic test for the patient. And then we focus on what are the implications for all other drugs."

Rule had been at Microsoft for a number of years, and decided that bioinformatics would be a logical place to focus, because the integration of computer science and biology would lead to fruitful and important endeavors. A trend in the personalized medicine space that Rule notes is that companies are focusing way too much on sequencing. However, he qualifies this by stating that, even though sequencing is important for the future, it is not ready yet. What Translational Software does is focus on relevant information that would change the way doctors treat their patients. By taking that information first, they are able to say that there are these 60 variants that are really important to prescribing medications, and you can come up with a PCR test that is cheaper. Getting this in doctors' hands sooner from a technology that is well proven today and matching it up with the clinical information that is also well proven today is a much more effective way to go forward. Their theory is that they will start offering information around these 60 variant panels, and they will grow with the industry because, frankly, there is an incredibly small amount of variation in the genome for which its importance is known. They are going gene by gene and drug by drug to build out the knowledge base. "There are some good online resources for what the guidance is. You need to go back to the literature, the basis of that."

Hachad refers to this literature as the basis for their recommendations. Hachad states:

> For pharmacogenetics, there are some consortiums that are in charge of reviewing the literature and making evidence-based recommendations. There is one particular consortium attached to the NIH. The consortium is called CPIC [Clinical Pharmacogenetics Implementation Consortium], and they have published a number of evidence-based guidelines, genes of interest, and drugs of interest that we have used to implement in our program. We are heavy users of these guidelines. The authors are experts in the use of pharmacogenetics, some of them publishing and making these studies for several years. We highly believe their assessments. In some cases, we do not have guidelines and when our customers are interested in providing gene–drug associations, we do our own assessment of the literature, but we label it differently. We label the evidence differently than we use CPIC guidelines. Because we want doctors to understand [the evidence-based differences].

For the simple translation of the genetic test result into the guidance, Rule and Hachad were surprised to learn how many of the tools for the conversion of a genetic test result from a haplotype to a phenotype were often focused on research and not really appropriate for clinical use. So, there is a whole range of things that are going to need to be built out to make this really useful. From a software perspective, an awful lot is focused on research, so it does a good job at looking at associations across populations, but it does not necessarily say for an individual patient what is the right approach. They see this as an area to move into. Integrating into the existing clinical workflow is really important. One sees a lot of people who put up a website and ask the doctor to come to their website. That is really difficult for someone who is really busy. So, figuring out how this works with their existing processes is really important. In the short term, what that means is that they will send in a paper requisition and get something back. It is all technology they already have. Where Rule sees the world going is that we will send that data into their medical record system and order and retrieve that information from the system, and he is really interested in working to take the genetic test result and putting it in the medical record. Again, this is 60 variants; it is not a lot of data right now. But they are coding that in such a way that the next time they prescribe a medicine, they can use both the genetic test result and drug list and see what the implications are. This is an area of active work right now.

> The guidance we have found is almost exclusively focused at researchers not at clinicians. So first of all, you have to look for mistakes, and then putting it in terms that a doctor of ordinary experience can understand. In some cases, we have to lower the sensitivity to get a better result, but that is far better than giving the doctor no answer.

Rule explains how their processes work. They market to the lab; it is expensive to get their products to the physician. The cost of their service is relatively small in the scheme of things. Rule notes that it makes more sense to sell to the lab and the lab reaches out to the physician because they are selling the whole entity. That has worked out very well. The gap that they fill is that the software the company develops is too complex for each lab to build itself. Rule and Hachad envision a tipping point in the future: rather than each lab building its own proprietary tools, they see them working with Translational Software to more effectively provide the interpretation.

Among the trends that Rule observes are that right now, people are using genetic testing in pharmacogenetics; in a lot of cases, they are reactive. Prescribe clopidogrel, and then you get a test to see if it will be effective. What Translational Software envisions over time is that at one point, you will have a panel of things tested and be able to record that for the rest of your life. So, the incremental value of using that data for individual prescriptions is that each of them may not necessarily justify a genetic test, but if one has that test done at some point, it becomes a resource that he or she can draw on for the rest of his or her life.

Rule describes how their reports become filtered for physician use. The other area where they are going is rather than providing static information of just a report, they are dynamic, enabling the doctor to drill down on that report and click on relevant information. "Keep that data updated," they state. The static report relays what medications a patient is taking at that time. A more dynamic capability is to say what drugs the patient is currently taking and how they compare with the current state of knowledge because the back-end knowledge base is evolving pretty quickly. That is complicated by the fact that the medical record market is pretty fractured. Often, clinical institutions will buy "best of breed," so they have several different systems, as opposed to integrated systems. Right now, any institution has a fairly large collection of different software that they use.

What Rule and Hachad find is that physicians are beginning to respond to personalized medicine. When you talk to a pain doctor who is going to prescribe a controlled substance to somebody in chronic pain, he or she is interested in knowing whether it is going to work. In areas of psychiatry, where it is not uncommon to go through two to four rounds of antipsychotics or antidepressants that might work, they are interested in being able to reduce the level of iteration. Translational Software is not at the point where they can tell doctors what will work, but they can eliminate some things that they know *a priori* will not work. Rule states:

I have to say the market is growing faster than I expected. I started the company in 2009, and honestly, if I had product now that I had in 2009, I don't know if I would [have] sold anything. But we do see huge amount of interest now because there is a general cultural perception of genomic medicine being something interesting and worthwhile. I think we are beginning to get past what Gardener calls the "trough of disillusionment." If you are familiar with the "hype cycle," we have gone through the hype, and people are just beginning to look around to see what works, and pharmacogenetics is probably going to be the first and most important implementation of personalized medicine because pharmacy affects everyone.

Hachad states:

If you compare pharmacogenomics with other types of personalized medicine, we have less ethical implications to think about because it is [a] pretty straightforward message that you get if you are testing for a gene that is affecting a drug response. And we are not going into predicting disease risks or incidental findings. When you compare with other applications of genomic medicine, where the ethics can be really a problem, where you need patient counseling to explain risks of variants of unknown significance.

We are only testing for things we know what to do with, that are actionable. We are looking for specific things where we know exactly what they mean.

Efforts into bringing diagnostics into the clinic have also emerged. Genomic Expression, a biotechnology start-up based in New York City, is another innovative company in the space of personalized medicine. Gitte Pedersen, CEO, notes how implementing biomarkers in the clinic is a lengthy affair, as shown Figure 10.2.

According to Pedersen, it can take 10–20 years to get a clinically relevant biomarker implemented in the clinic, while the number of clinically relevant biomarkers is increasing rapidly (Figure 10.2). The technology that Genomic Expression harbors uses RNA as the candidate marker species for next-generation diagnostics. The advantages are that

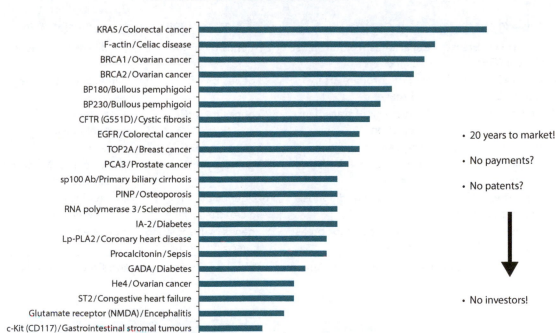

Figure 10.2 Graph illustrating for each disease the number of years from discovery to the FDA-reviewed test. It takes approximately 20 years to market. (Courtesy of Gitte Pedersen, 2015 Arrowhead Publishers Personalized and Precision Medicine Conference.)

RNA resembles the phenotype of the cell, doctors can easily translate results into clinical use without additional genetic education, first-generation tests are currently reimbursed in large markets such as cancer and cardiology, and diagnostic algorithms based on RNA are patentable, unlike DNA biomarkers.

The following are examples of multigene expression diagnostic assays costing from $3500 to $6000 (Table 10.3).

Pederson claims that Genomic Expression's technology can deliver all these algorithms through its proprietary OneRNA sequencing assay, which can identify and quantify all transcripts (Figure 10.3).

Pederson asserts the need for this technology by noting the following statistics:

- Every year, the U.S. healthcare system spends about $80 billion on cancer drugs while 8 million patients die.

- Only one out of four cancer treatments prolongs life.
- Only 10% of early breast cancer patients benefit from chemotherapy.
- Lung cancer patients with ALK mutations can live three years if treated with Xalkori, instead of six months of chemotherapy.
- In 95% of cases, acute lymphoblastic leukemia (ALL) is caused by a specific translocation or fusion and treatment is highly effective.

Pedersen also claims that immunotherapies such as Keytruda are the only class of drugs that may deliver prolonged disease progression, and yet they can cost $150,000 per year.

Of the three types of treatment options, chemotherapy, targeted therapy, and immune therapy, DNA panels can predict response to targeted therapy (30% of the market), while RNA is able to predict response to all three. OneRNA is purported to

Table 10.3 Diagnostic tests and use and type of technology. qPCR, quantitative PCR

Test (developer)	Use	Signature type	Regulatory/ reimbursement
Oncotype Dx Breast Cancer (GH)	Recurrence risk for breast cancer	21-genes by qPCR	Medicare reimbursed
Corus CAD (CardioDx)	Risk of coronary artery disease	23 genes by qPCR	Medicare reimbursed
AllopMap (XDx)	Predict risk of transplant rejection	20 genes by qPCR	FDA cleared, reimbursed
Pervenio (LifeTech)	Lung cancer prognostic	14 genes by qPCR	
Mammaprint (Agendia)	Recurrence risk for breast cancer	70 genes by microarray	FDA cleared, Medicare reimbursed
Afiirma Thyroid FNA Analysis (Vericyte)	Thyroid cancer diagnosis	142 genes by microarray	Private insurer reimbursed
Prosigna (Nanostring)	Recurrence risk for breast cancer	50 genes by Nanostring barcoding method	FDA cleared

Source: Courtesy of Gitte Pedersen, 2015 Arrowhead Publishers Personalized and Precision Medicine Conference.

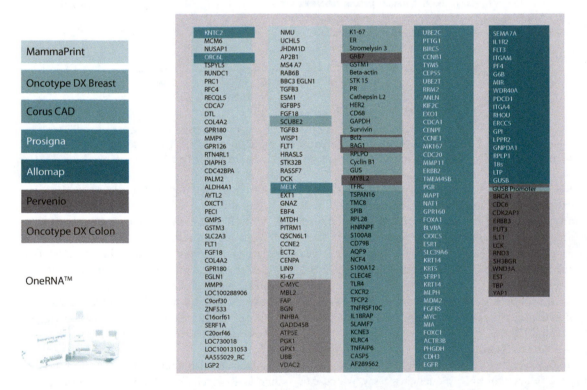

Figure 10.3 OneRNA can replace all these assays. (Courtesy of Gitte Pedersen, 2015 Arrowhead Publishers Personalized and Precision Medicine Conference.)

go from RNA to actionable report. Currently, its proprietary sequence algorithms produce databases with all actionable RNA targets in oncology, and Genomic Expression has a strategic partnership with IBM delivering OneRNA analysis in a Health Insurance Portability and Accountability Act (HIPAA) cloud. OneRNA can produce reports such as those for triple-negative breast cancer, which has

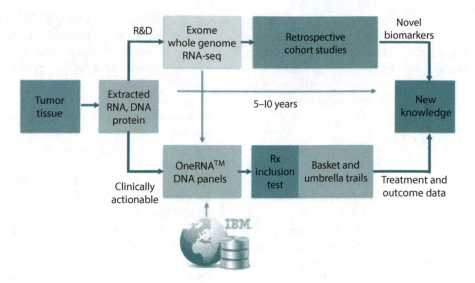

Figure 10.4 OneRNA workflow. Tumor tissue is taken, RNA is extracted, and OneRNA panels are performed, followed by clinical trials. (Courtesy of Gitte Pedersen, 2015 Arrowhead Publishers Personalized and Precision Medicine Conference.)

very few options and poor prognosis. OneRNA identified approved targeted and immune therapies and many clinical trials that this patient could be eligible for. OneRNA also boasts of fast adoption of actionable biomarkers by updating the cloud. In contrast to other RNA platforms utilizing panels, the OneRNA assay remains the same; thus, when a new expression marker has been identified to be clinically relevant, the only change required to implement it into clinical practice is updating the cloud-based software (Figure 10.4).

Pedersen also claims that OneRNA can revolutionize the clinical trial process. She describes the current protocol for double-blind randomized trials, specifically the umbrella trial, as shown in Figure 10.5.

Pederson has a vision for OneRNA that the tumor profiles it generates would match to clinical trials where compressive detection of cancer drugs targets in tumor using cutting-edge technologies takes place, resulting in curated databases of clinical trials and reports linked to expression biomarkers. This

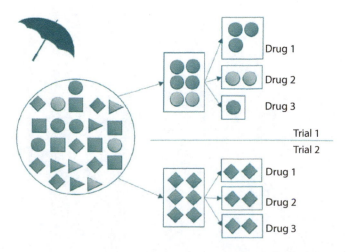

Figure 10.5 Umbrella trial for OneRNA. Targeted therapies would be recategorized in the trial according to the diagnostic panel. (Courtesy of Gitte Pedersen, 2015 Arrowhead Publishers Personalized and Precision Medicine Conference.)

would assist pharma in bringing the best patient for the drug. As she states, "By leveraging a partnering network, we can add outcome data to develop robust responder algorithms in collaboration with Rx companies and deliver them through our platform in real time to our partnering institutions."

CONCLUSIONS

From CLIPMERGE to BioVU to ClinGen, various personalized medicine centers around the country are furthering the aims of precision medicine through their projects to bring research data into the clinic, enhancing clinical utility. Actionable reports from companies like Translational Software and Genomic Expression bring pharmacogenomics data and diagnostic profiling to clinicians in crisp fashion, enabling the implementation of personalized medicine from biomarker discovery to clinical outcomes.

REFERENCES

Baranov, V.S. 2009. Genome paths: A way to personalized and predictive medicine. *Acta Naturae* 3:70–80.

Colley, K.J., Wolfert, R.L., and M.E. Cobble. 2011. Lipoprotein associated phospholipase A(2): Role in atherosclerosis and utility as a biomarker for cardiovascular risk. *EPMA Journal* 2:27–38.

Krol, A. 2013. The druggable genome is now Googleable. *Bio-IT World*, November 22. Available from http://www.bio-itworld.com/2013/11/22/druggable-genome-now-googleable.html.

Kumar, P. et al. 2013. Circulating miRNA biomarkers for Alzheimer's disease. *PLoS One* 8:7.

McKinnon, R.A., Ward, M.B., and M.J. Sorich. 2007. A critical analysis of barriers to the clinical implementation of pharmacogenomics. *Therapeutics and Clinical Risk Management* 3:751–759.

Walko, C.M., and H. McLeod. 2009. Pharmacogenomic progress in individualized dosing of key drugs for cancer patients. *Nature Clinical Practice Oncology* 6:153–162.

CHAPTER 11

The economics of personalized medicine

Price is what you pay, value is what you get.

Clifford Hudis, MD
Memorial Sloan Kettering Cancer Center

In the United States, 18% of the GDP is being spent on healthcare. In 2020, this figure is projected to rise to 20%. However, this higher spending does not increase life expectancy or lead to better health outcomes. Healthcare reform occurred, leading to better access to care. Cancer has been somewhat of a success story, but now more people are living with cancer: 42% of the population will live with cancer in the near term.

How does personalized medicine fit into this economic picture of healthcare? How can the value of companion diagnostics be determined, and what about the price of personalized drugs and access to them? What role do insurers and the pharmaceutical industry play?

This chapter sheds some light on these questions through the insights of health economists and players in the pharma field. Personalized medicine will definitely have an impact on the economy, perhaps positive. However, whether evidence-based outcomes will prompt payers to reimburse for expensive personalized oncology drugs or whether pharma will invest in companion diagnostic codevelopment remains to be seen.

Centers for personalized medicine are emerging around the country and stimulating local economies, as a recent study by the Mount Sinai School of Medicine reported. According to a 2008 PriceWaterhouse Coopers report, personalized medicine is expected to have a significant impact on the economy, as well as on healthcare costs, with the impact on costs differing in the short term versus long term, and personalized medicine is a disruptive innovation that will up-end traditional business models, create new economic models and funds flow, and reallocate healthcare resources away from disease treatment and toward wellness and prevention. Personalized medicine is becoming the new paradigm in healthcare that is transforming the way $2.6 trillion is spent in the United States and $4 trillion is spent globally (McDougall, 2008; PriceWaterhouse Coopers, 2008).

In review, what is "personalized medicine"? It is cures and therapies designed through strategies that tailor-make medicines for a person's DNA to treat disease, prevent illness, and maintain health. A $5 billion market, the personalized medicine market will double in five years from 2009 to 2014. Associated health markets are estimated at $400–$900 billion. The players in the healthcare field (payers, providers, pharma, and biotech) are advancing progressively in this market through a number of activities, including research and development, investments, and mergers. Many companies outside the healthcare field, such as Google, Microsoft, Cisco, Intel, Oracle, and Walmart, are beginning to operate in this paradigm, and will benefit considerably.

Personalized medicine promises to increase health value (maximize health outcomes for every dollar spent on healthcare) by increasing innovation through the entrance of more companies, and advance preventive medicine by enabling precise interventions against disease.

Personalized medicine is expected to have a significant impact on the economy as well as on costs, with the impact on costs differing in the short-term versus long-term. In the long-term, personalized medicine will likely decrease the costs of healthcare because of:

- The emphasis on prevention, which will reduce the incidence of disease and related costs
- The development of improved diagnostics, which will eliminate the costs associated with:
  - Validating positive diagnostic tests (e.g., through expensive imaging procedures)
  - Implementing treatments that produce no benefits and may even cause harm
- A focus on early detection and ongoing monitoring, which will eliminate the need for the expensive procedures, devices, and drugs required by current health interventions, many of which target late-stage disease.

In addition, gains can be expected due to the increased productivity and reduced absenteeism resulting from the improved overall health of the workforce (PriceWaterhouse Coopers, 2008).

As noted in a Chapter 2, personalized medicine is also known as stratified medicine. As Trusheim et al. (2007) write, most medicines are currently prescribed based on symptomology. Citing nonsteroidal anti-inflammatory drugs (NSAIDs) for pain relief, proton pump inhibitors (PPIs) for gastroesophageal reflux disease (GERD), and the Gardasil vaccine for human papillomavirus (HPV) infections, they note that drugs are prescribed by clinicians and tested on a trial-and-error basis. They also note

advances in medicine that match selective groups of patients with safe and effective therapies. Calling this stratified medicine, Trusheim et al. (2007) pose the category of clinical biomarkers, which enable the targeting of patients with treatments.

Gleevec is described as an example of stratified medicine, The BCR-ABL tyrosine kinase genotype is a clinical biomarker that can identify chronic myeloid leukemia (CML) patients who would most likely respond to Gleevec. Stratified or personalized medicine will thus change the paradigm of forging anticancer therapies.

However, this paradigm shift in biomedicine will be confronted with challenges. First and foremost, many patients remain wary of personalized medicine and its accompanying race-based genomics. Issues of genetic discrimination for personalized medicine testing abound, and patients are worried that their genetic information may be used against them by insurance companies and employers. Insurance companies may not be willing to pay for genetic testing and diagnoses or the targeted therapy being developed for patients. Pharmaceutical companies, built on the premise of the blockbuster drug or one-size-fits-all drugs, may not get onboard with this new paradigm of drug development. Furthermore, there are also scientific impediments, such as a possible dearth of molecular markers to diagnose disease and make targeted therapies, that may obstruct personalized medicine. The information technology sector may not keep up with personalized medicine initiatives. Personalized therapeutics is also thought to be a threat to health equity (Ward, 2012). However, as a 2010 report by the business consultancy company McKinsey & Company suggests, the main hurdles are economic. Success will require planning robust prospective trials, analyzing healthcare economic and outcome data, assuaging insurance and privacy concerns, developing health delivery models that are commercially viable, and scaling up to meet the needs of the whole population (McKinsey & Company, 2010).

As Waldman and Terzik (2012) write:

We are in the midst of a revolution in disease management established by emerging innovations in platform technologies, where clinical outcomes are resolved by targeted molecular diagnostics and therapeutics. This evolving paradigm has already yielded products that have advanced into the healthcare marketplace. In the context of world-wide economic realities and constrained healthcare resources, it is essential to establish the value proposition of targeted diagnostics and therapeutics, to ensure their benefits are maximized for patients, populations, and societies.

Personalized medicine is showing new economic models for the biopharmaceutical industry. Personalized medicines cost more per patient, however, because they lead to better outcomes in a targeted group; personalized medicine drugs can accrue greater monetary gains. An example is shown by Gleevec. Annual revenues for Gleevec thus far are $2.5 billion. At a cost of $43,000 per patient, with 55,000 patients receiving the drug, Gleevec can be compared with tamoxifen (a therapy for breast cancer), which generated $650 million in annual revenues with 500,000 patients. Personalized medicines can be considered superior to traditional medicines through their efficient targeting. Figure 11.1 shows three distinct $1

billion product regions: classic blockbusters, high-value orphan drugs, and the emerging targeted or stratified medicines (sometimes called "niche busters").

Economic factors, of course, prominently figure into the implementation of new technologies. A positive return on investment (an investment's rate of return exceeds an investor's expected rate of return) is usually the indicator to invest in novel research and development projects. The issue of availability of resources also impacts the adoption of a technology. However, as Paci and Ibaretta (2009) write, health-related sectors have some peculiarities that create a more complex picture. These markets involve complex transactions among several different competing and cooperating actors: producers, public and private payers, regulatory agencies with diverse structures, patients, and physicians. (As Table 11.1 indicates, each of these actors has differing perceptions of personalized medicine.)

As suggested by Davis et al. (2010), a major barrier to the diffusion of a new technology is the fact that it might not be economically attractive to all

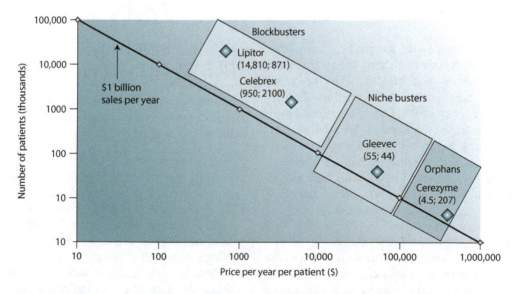

Figure 11.1 Different types of high-revenue medicines. Classic blockbusters such as atorvastatin (Lipitor) are prescribed to large patient populations. Stratified medicines that can be demonstrated to be clearly superior to alternatives for a smaller patient population can also achieve high revenues, resulting from higher pricing and adoption rates. At one extreme are orphan drugs, which often represent the only therapy for a small population with the disease in question, and which can therefore support even higher pricing if they are highly effective and the condition is sufficiently severe. The numbers in brackets correspond to the number of patients (thousands) and the price per year per patient ($), respectively. (From Trusheim, M.R., Berndt, E.R., and F.L. Douglas. 2007. *Nature Reviews Drug Discovery* 6:287–293.)

Table 11.1 Physician, patient, industry and payer perceptions of personalized medicine

The perception of personalized medicine varies between key stakeholders, with all stakeholders considering the value for money and cost of new treatments a central decision criterion		
Physicians/patients	Industry	Payers
• Physicians welcome the availability of additional treatment options that offer improved efficacy and safety • Patients hope for curative treatments that improve the health status • Payers and physicians are often disappointed about the incremental benefit and cost of personalized medicine • Perceived with great aspiration but promise not yet fulfilled	• Perceived as major growth area for the future • Commercial exploitation of new and promising treatment concept • Improve the success rate in development • Accelerate regulatory and payer approval and extend on-patent • Personalized medicine is an attractive area that may allow to enter into a promising and rewarding new area of sciences and business	• Perceived with high ambiguity because the implications are not yet clear • Hope for savings resulting from targeted administration of drugs • Concern that new cost that may exceed the saving potential • Potential to change the established value for money ratio in health care • Affordability is the key priority of payers; can I afford all the good things coming along with the budget that is available?

Source: From Caesar, M., Pharmaco-economic aspects of personalized medicine and companion diagnostics. Easton Associates.

stakeholders. The public sector in healthcare often has a strong influence in shaping the adoption and innovation of new technologies. Given the limitation of healthcare resources and the limited capacity of governments to finance all desired health interventions, the evaluation of potentially expensive emerging health technologies such as pharmacogenetics and pharmacogenomics is crucial. Economic evaluations of new health technologies are increasingly used to inform policy makers of which interventions can or should be introduced in the delivery of healthcare.

Personalized medicine, and associated pharmacogenetics tests, can lead to innovation across the healthcare spectrum. Pharmacogenetics drugs can improve patient benefits by reducing costly adverse drug reactions. Pharmacogenetics can also lead to greater understanding of disease mechanisms and drug responses, leading to further improvement in drug development by increasing speed and decreasing cost. From biomarker discovery to public health, cost-effectiveness can enable the implementation of personalized medicine (Paci and Ibaretta, 2009).

ESTABLISHING VALUE FOR MOLECULAR DIAGNOSTIC TESTS

The value-based approach is needed in economic models in predictive, preventive, and personalized medicine (Akhmetov and Bubnov, 2015).

Fundamental shifts in the healthcare paradigm, driven by the gradual transition to "patient-centered" value-based health service delivery, open new horizons to personalized medicine offering the potential to provide timely and cost-effective medical solutions to stratified patient subpopulations with predictable outcome margins. By using biomarkers as measurable indicators of predisposition to or severity of a disease state, personalized medicine helps in early detection, monitoring, assessment of risks associated with a disease, and guiding therapeutic decisions. Unlike therapeutics (Rx), which undergo three phases of clinical trials prior to marketing authorization and whose effect on patients can be quite straightforwardly demonstrated by patient-reported outcome measures, there is no clarity in the molecular diagnostics field on how much evidence is required to prove the value of a test. As a rule, most of diagnostic studies focus on accuracy and feasibility, which is a hardly enough prerequisite for better patient health or other downstream improvements. High value of molecular diagnostic testing is observed in the pharmaceutical pipeline, as it facilitates discovery of biomarker-based therapies targeting disease causes instead of symptoms. It is evident that today, 50% of all clinical trials conducted by pharmaceutical companies collect DNA from patients in order to facilitate biomarker development. Moreover, is also known that biomarker-based diagnostics used in clinical trials can increase chances of regulatory approval and enhance prescription (Akhmetov and Bubnov, 2015).

Porter's value-based healthcare (VBH) model places patients atop of the hierarchical pyramid of significance, arguing that the true value of any healthcare service, including diagnostic testing, can only be conceived through a prism of the total bundle of products and services delivered to an individual patient over a cycle of care by correlating the final patient outcomes with the associated costs. In other words, the only way to adequately assess the real value of a diagnostic test is to consider it an integral part of the combined "efforts" of all links in the value chain involved in the process of delivering value to a patient (e.g., healthcare providers, caregivers, manufacturers, pharmacists, and laboratories), assigning weights to each link based on its contribution to the overall patient outcome, and measuring the total costs required to deliver this outcome:

$$\text{Value} = \frac{\text{Patient outcomes}}{\text{Costs of delivering the outcomes}}$$

The patient outcomes in this model include prevention of illness; early detection; right diagnosis; right treatment to the right patient; rapid cycle time of diagnosis and treatment; treatment earlier in the causal chain of disease; less invasive treatment methods; fewer complications; fewer mistakes and repeats in treatment; faster recovery; more complete recovery; greater functionality and less need for long-term care; fewer recurrences, relapses, flare-ups, or acute episodes; reduced need for emergency room visits; slower disease progression; and fewer care-induced illnesses.

PAYER PERSPECTIVE

In the recent survey among small and medium businesses dealing with health technology assessment (HTA), the interviewed firms accepted that the main impediments in the reimbursement process for their products were poor understanding of specific payer requirements, insufficient scientific advices from the HTA bodies, lack of methodological agility and unnecessary bureaucracy resulting in belated identification of genetic variants, inadequate description of clinical trial designs, incomplete representation of patient experiences, etc. Additionally, the review of scientific articles containing evidence to help guide decision-making about insurance coverage for Alzheimer's disease diagnostic tests, conducted by the Institute for Clinical and Economic Review (ICER) and Policy Development Group (PDG), showed that all studies without exclusion failed to provide convincing evidence that payers could use to showcase improved outcomes. None of these studies established analytic validity by capturing action based upon diagnosis, health outcomes (e.g., cognitive/function decline), societal outcomes (e.g., cost-effectiveness), or technical efficacy. As a result, a great number of companion MDx developed separately from therapeutics did not receive wide payer acceptance due to poorly established links between testing, therapeutic interventions, and health outcomes (e.g., tests to estimate warfarin dosage or CYP2C19 assays to stratify clopidogrel-eligible subpopulation). Additionally, according to the literature review by Paci and Ibaretta, 27%

Table 11.2 Coverage inconsistencies in molecular diagnostics

Innovative test example	Positive coverage policies			
	Aetna	Regional CMS	Cigna	Regional BCBS
AlloMap™		X		
OncoType Dx™ (breast cancer)	X	X	X	X
MammaPrint™		X		
BRACAnalysis™	X	X	X	X
OVA1™		X		X
KRAS (ovarian cancer)	X	X	X	X

Source: Reprinted with permission from Akhmetov, I., and R.V. Bubnov. 2015. Assessing value of innovative molecular diagnostic tests in the concept of predictive, preventive, and personalized medicine. *EPMA Journal* 6:1–12.

of the assessed Dx tests failed to demonstrate favorable and univocal cost-effectiveness evidence compared to the standard of care (Table 11.2) (Akhmetov and Bubnov, 2015).

In an economic model developed by Trusheim and Berndt (2015), setting the companion diagnostic performance by selecting the cutoff score integrates scientific, clinical, ethical, and commercial considerations and determines the value of the stratified medicine combination to developers, payers, and patients.

The stratified medicine companion diagnostic (CDx) cut-off decision integrates scientific, clinical, ethical, and commercial considerations, and determines its value to developers, providers, payers, and patients. Competition already sharpens these issues in oncology, and might soon do the same for emerging stratified medicines in autoimmune, cardiovascular, neurodegenerative, respiratory, and other conditions. Of 53 oncology targets with a launched therapeutic, 44 have competing therapeutics. Only 12 of 141 Phase III candidates addressing new targets face no competition. CDx choices might alter competitive positions and reimbursement. Under current diagnostic incentives, payers see novel stratified medicines that improve public health and increase costs, but do not observe companion diagnostics for legacy treatments that would reduce

costs. It would be in the interests of payers to rediscover their heritage of direct investment in diagnostic development (Trusheim and Berndt, 2015).

A stratified medicine competition is shown in Figure 11.2. To illustrate implications of stratified medicine competition, consider the hypothetical but plausible situation of three oncology candidate therapies in a race to be first and best in class to treat the same novel target for which a candidate companion diagnostic for likely drug responders has been identified. For this example, we assume that the three candidate drugs are essentially similar in their chemical structure, pharmacology, formulation, therapeutic index, and other relevant properties. This enables us to focus on the decisions and implications regarding whether and how to use the candidate companion diagnostic. Figure 11.2 demonstrates three possible choices facing the firms based on these assumptions. Firm A chooses an all-comers approach for its Drug A, which does not use the companion diagnostic. Firms B and C both choose to pursue a companion diagnostic approach, but set different cutoff values. Firm B chooses a low diagnostic cutoff value for its Drug B, whereas Firm C sets a high diagnostic cutoff for its Drug C. As Firm A anticipates launching Drug A, it faces the isorevenue, depicted by the curve with boxes, in Figure 11.2b. The all-comers

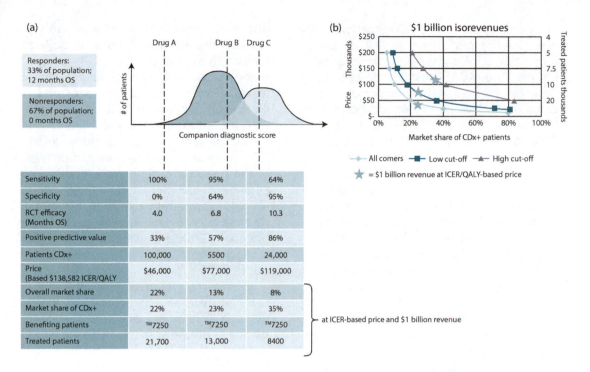

(a)

Drug A   Drug B Drug C

Responders:
33% of population;
12 months OS

Nonresponders:
67% of population;
0 months OS

# of patients

Companion diagnostic score

(b)

$1 billion isorevenues

Market share of CDx+ patients

—◆— All comers —■— Low cut-off —▲— High cut-off
★ = $1 billion revenue at ICER/QALY-based price

	Drug A	Drug B	Drug C
Sensitivity	100%	95%	64%
Specificity	0%	64%	95%
RCT efficacy (Months OS)	4.0	6.8	10.3
Positive predictive value	33%	57%	86%
Patients CDx+	100,000	5500	24,000
Price (Based $138,582 ICER/QALY	$46,000	$77,000	$119,000
Overall market share	22%	13%	8%
Market share of CDx+	22%	23%	35%
Benefiting patients	™7250	™7250	™7250
Treated patients	21,700	13,000	8400

at ICER-based price and $1 billion revenue

Figure 11.2 The economic possibilities for three drugs with companion diagnostics on the market. (Reprinted with permission from Trusheim, M.R., and E.R. Berndt. 2015. *Drug Discovery Today* 20:1439–1450.)

label supported by its clinical trial will allow its marketing to suggest that all 100,000 patients with the condition are eligible for treatment. To achieve $1 billion blockbuster-level sales, Drug A must achieve 20% market share (be used by 20,000 patients) at a $50,000 one-year drug regimen price with its four-month overall survival improvement. Firm A could choose any other price, and the isorevenue, depicted by the curve with boxes, indicates what market share Drug A must achieve for $1 billion in sales. For instance, at a price of $200,000, Drug A must be used, and paid for, by 5000 patients, which conveniently equals 5% market share in this example. At a price of $12,500, 80% market share must be achieved (80,000 treated and paid patients) to reach $1 billion in sales. A recent literature review suggests that for oncology therapeutics in the United States, the mean incremental cost-effectiveness ratio (ICER) using the quality-adjusted life year (QALY) for the

health benefit metric is $138,582/QALY. If the ICER guides payer reimbursement, Firm A might expect a Drug A price of approximately $46,000 (one-third of the mean ICER based on an expected average four-month overall survival improvement). Per Figure 11.2b, at that price, Drug A would need to achieve 22% market share to generate $1 billion in annual revenue, as indicated by the star.

Firm B chooses to use a companion diagnostic approach that selects nearly all patients who will respond by setting a low CDx cutoff. In this hypothetical case, the cutoff is set to generate 95% sensitivity (95% of responders will receive a positive test score: approximately 31,500 of 33,000). The hypothesized test is assumed to be good, but not perfect. The low cutoff value results in a 64% specificity (64% of nonresponders will test negative: approximately 43,000 of 67,000). This means that 36% of the nonresponders will test positive (approximately 24,000 of 67,000). For an oncology companion diagnostic, this is a superior performance. One of the more powerful companion diagnostics known, the KRAS test for detecting likely responders and nonresponders to cetuximab (Erbitux) in colorectal cancer has an estimated 75% sensitivity and 35% specificity. Firm C chooses to use a companion diagnostic approach that excludes nearly all patients who will not respond by setting a high diagnostic cutoff. In this hypothetical case, the cutoff is set to generate 95% specificity (95% of nonresponders [approximately 63,500 of 67,000] will receive a negative test score [CDx–]). As shown by the far right vertical line in Figure 11.2a, the high cutoff also excludes some patients who would benefit from treatment. In this hypothetical case, the corresponding sensitivity is 64% (approximately 21,000 of 33,000 patients who would respond); 36% of patients who might benefit (approximately 12,000 of those patients).

A savvy payer or integrated provider might recognize that the drugs are identical and therefore use the CDx with Drug A (or negotiate discounts with Firms B and C to match Drug A pricing). Using the low cutoff to reach nearly all responders but with Drug A pricing would lower the ICER-based price to approximately $81,000 and the total cost to $2.6 billion to achieve the nearly perfect health benefit of approximately 32,000 QALY/year and save the payer $1.7 billion—over 35% compared with the ICER-justified Drug B price. Such actions would of course reduce incentives for future developers to develop stratified medicines if in the end they still only receive the all-comers nonstratified value.

The companion diagnostic developer faces lower revenue prospects. Even assuming a high reimbursement to the clinical laboratory of $400, of which the CDx developer receives 50% for the test kit, the entire CDx testing market is only $20 million (for a selection, but not monitoring CDx) compared with a market measured in billions of dollars for the therapeutics. If the CDx uses a standard technology, such as an immunoassay whose kit prices are often effectively limited to $25 or less per test, the total market falls to merely $2.5 million. Given rapid competition, most CDx developers will receive half or less of the potential testing market, and not all patients will be tested, making these already comparatively small amounts even smaller. Even a manufacturer test price of $2000 per patient with 100% testing captured by the firm only produces $200 million per year.

In summary, under plausible market competition characteristics, three essentially identical drugs receive dramatically different labels, ICER-justified pricing, and market positioning, whereas under all circumstances, the CDx developer likely receives 1% or less of the revenue flowing to the therapeutic. Stratified medicines tend to focus on significant unmet needs, as indicated by their disproportionate priority review designation and qualification for accelerated or breakthrough medicine approval. Payers, and the governments and employers who fund them, now face the challenges of paying for the emerging successes in meeting those unmet medical needs. Unsurprisingly, incremental improved public health is likely to have incremental costs, even if a stratified medicine approach proves reasonably efficient at identifying those who will benefit (Trusheim and Berndt, 2015).

They conclude

Stratified medicine tightens the links among science, the clinic, and the marketplace. Setting the companion diagnostic cut-off value is the crucial shared connection among all three, with no easy rule of thumb to guide the choice. Each stratified medicine opportunity faces unique facts and circumstances that require balancing ethical, scientific, and financial concerns. Today, stratified medicine economic incentives favor new medicine developers and the patients they serve. Payers benefit from more efficient new treatment for unmet medical needs, but likely face increased total costs for the resultant increase in overall public health, with little or no offsetting cost savings from companion diagnostics better stratifying legacy treatments. Diagnostic companies are generally paid for their services, but not sufficiently to invest independently in companion diagnostic development. Current economics do not reliably signal true healthcare needs to therapeutic and diagnostic developers, and even less so to the discovery scientists at the beginning of the innovation value chain (Trusheim and Berndt, 2015).

PERSONALIZED MEDICINE AND PHARMACOGENOMICS: FROM MICRO TO MACRO

According to a report on the economics of personalized medicine by Easton Associates, the promise of personalized medicine is to improve healthcare by increased treatment effectiveness with minimal side effects, and to realize savings potential by eliminating the cost associated with treatment failure and adverse drug events (Caesar). In terms of assessing the macroeconomic impact of personalized medicine, the overall effect on the economy, microeconomic factors must be indicated in the form of cost-effectiveness, cost utility, and clinical health outcomes. According to Wong et al. (2010), questions to consider in assessing the cost-effectiveness of a pharmacogenomics treatment strategy include the frequency of the genetic variation and its association with a drug response, environmental factors of a drug response, the sensitivity and specificity of the genomic test, outcomes of the treatment, and the effectiveness of monitoring strategies for preventing adverse drug reactions and predicting drug response. Wong et al. (2010) conclude that for pharmacogenomics to be cost-effective, the genetic variation must be prevalent, the genetic testing for that variant must be highly

sensitive and specific (few false positives and false negatives), and the treatment must involve significant outcomes or costs that are impacted by personalized therapy.

Cost-effectiveness provides evidence comparing costs and health outcomes of corresponding health technologies. Targeted therapies, like genetic screening for newborns, can prove to be cost-effective for payers based on their indication for disease risk and drug response.

In a McKinsey & Company (2010) report on the microeconomics of personalized medicine, Davis et al. conclude the poorly aligned economic incentives among stakeholders could serve as an impediment to the implementation of personalized medicine. Even though certain genetic tests can help avoid possibly expensive and inappropriate therapies, minimize costly adverse events (such as the warfarin dosing test), or at least delay expensive yet useful procedures, payers such as insurance companies are slow to invest in personalized medicine. Providers face both microeconomic incentives and disincentives to use personalized medicine tests. Through the development of biomarkers (an indicator of a biological state), pharmaceutical and biotechnology companies are investing in companion diagnostics—tests to identify a patient's likelihood of responding to a drug or experiencing side effects—at various levels. A Brookings Institute report concluded that "our analyses show that integrating genetic testing into (therapy) significantly improves health outcomes and reduces healthcare costs" (Davis et al., 2010).

Robert Dvorak, PhD, currently a consultant and former information technology director of Genitope, a personalized medicine company that went out of business in 2008 and used to make monoclonal antibodies to treat each individual's cancer (made-to-order drugs), comments on the economics of personalized medicine, and makes the distinction between "made-to-order" and "made-to-stock" drugs:

In terms of economics, there is a huge difference in terms of talking about targeted therapies, where I am using genomic analysis to determine which monoclonal antibody you might react to appropriately. I have got a pool of five to six different monoclonal antibodies that might be used for breast cancer treatment. By looking at your particular surface proteins, I might be able to come up with a cocktail that [is] best suited to attack your particular cancer. In most cases, we are looking at [a] make-to-stock drug, and all I [am] doing is making a determination that you would do better with monoclonal A versus monoclonal B. There is no reason that you couldn't apply this to small-molecule drugs, to be able to say you know that you are one of people due to your genomic makeup [who] will not respond to Lipitor. I am not going to prescribe Lipitor for you, I am going to prescribe Crestor.

As Dvorak explains, Big Pharma is mainly invested in this type of personalized medicine, made-to-stock drugs. These kinds of things make economic sense to Big Pharma because it allows them to continue make-to-stock practices and also potentially allows them not to have a high degree of failure in their clinical trials, which comes from pure randomization. "To have my patient population for treatment be those people who will respond to the drug, it would radically change statistical work needed for drug approval." So, there is an advantage. The FDA has two criteria; one is for the whole diagnostics area, like 23andMe. The second, the whole school of personalized medicine that is not made to stock, is made to order. Potentially, a tissue is taken from a patient and a drug is made specifically for that patient, or it is returned somewhat adulterated, or perhaps some measurements or scans are taken from a patient and a medical device is made for him or her. One of the earliest forms of personalized medicine was personalized hip replacements, where the clinicians built a hip replacement for you, rather than just taking a random one and putting it in you. Those things have a different requirement.

Dvorak describes made-to-order drugs and the economics behind them:

First thing, they are made to order, which changes the economic issues. The second thing is that tissue taken from the patient belongs to the patient. One has a certain obligation to track the entire biopsy and return it or destroy it. So there is this extra burden of that tissue that [pertains] to the economics. But the biggest issue is if one makes a tissue for you, as a researcher at Genitope, and the patient has non-Hodgkin's lymphoma, and it's a nine-month process to take the biopsy from the patient and make the treatment. Let's say the patient decides that they do not want the treatment anymore. What does the company do with all the money that they have invested in the patient for this treatment? How will [the] company be reimbursed for a treatment that the patient will never receive? It changes the market for the field, which is why the personalized medicine field does not see investment by Big Pharma in that form of personalized medicine, which is arguably the single most transformative aspect of personalized medicine because it is personalized therapy as opposed to a targeted therapy.

But as Dvorak states, it has a huge economic burden. He explains the significance of made-to-order drugs.

It's [about] made-to-order [drugs]. It's nine months between the time the patient completes their biopsy and the time the patient completes their treatment. Is the company waiting nine months to get paid? The company is going to carry the entire debt burden of the patient's treatment regimen, with the case of Dendreon [a personalized medicine company that manufactures Provenge, a made-to-order drug], a patient getting treated with prostate cancer with their drug, Provenge [a personalized prostate cancer treatment] is paying $93,000—who is holding that $93,000 debt until the insurance company reimburses? Is it the maker of the drug, the distributor, who is carrying that debt? Because it's not going to be the doctor. No doctor out there is going to carry hundreds of thousands of dollars of debt for patients. And insurance companies are not going to pay until treatment is delivered. So there is a complex business model that goes with it; as such, it's going to raise a certain set of complications, another part of which is if somebody is paying $93,000 to make a treatment and no one is going to use [it] and you are throwing it away. There is a whole bunch of potential waste in our medical system, if people don't go through treatment, but people cannot be forced to go through treatment. New diagnostics might come along, their life conditions may change, more important medical issues may come up. So there are all sorts of complications that arise with this type of personalized medicine.

Dvorak explains the identity issues involved with made-to-order drugs:

What Dendreon does with Provenge is they take blood from a patient, they introduce into it an antibody to fight the prostate cancer, and they take that and inject it back into the person's body. Genitope took an antigen and created an immunostimulant and injected it into the patient's body so it makes antibodies to it. So there are different models. All of them come down to the company is making something specific for the patient. The costs are absorbed specifically for the patient's care and there are also some questions, in the case of Provenge, which are also the carrying costs, but Provenge extends mean time to progression by about six months. Should Dendreon be spending $93,000 to give a patient an additional six-month mean time to progression? Is that how society should be spending our health money? So there is a series of economic decisions here.

Dvorak notes:

Also let's say at Genitope, we messed up, and it wasn't your surface proteins,

we mixed you up with another patient, and we sent the treatment to you, what would happen is it would not work. It wouldn't hurt you, but it would not help you since there is not a cancer in your body that would react to that antigen. The flip side of that is there are drugs where if I took somebody's blood sample, and inject antibodies (Dendreon), then I could do some serious health damage to that person because I am using somebody else's blood, which may or may not be compatible with our own. So there are safety issues tied to identity. It runs you into conflict with HIPAA [Health Insurance Portability and Accountability Act] requirements. What patient has this disease is getting this treatment? Who is unblinding this patient to ensure correct identity? And what overhead is required to ensure those identity checks? So that comes back to the question of Big Pharma. So basically Big Pharma is happy to do diagnostic-based processes where they would be able to use the diagnostics for targeted therapy. It's relatively low risk and it's still make to stock.

Dvorak notes that Big Pharma is also interested in taking a drug that has a relatively low serving population and not having to worry about the economies of producing this drug, because they are not actually competing against something else. If they are making a made-to-order drug, they can be in big trouble economically.

So the drug we were doing at Genitope was for non-Hodgkin's lymphoma. The standard of care for non-Hodgkin's lymphoma is Rituxan, which is made-to-stock drug from Genentech. Arguably, Genitope's drug is better, but if you decide as an individual patient you are not taking Rituxan, Genentech can sell that drug to somebody else. You do not have a lot of risk. But if you choose to do Genitope's therapy and decide not to go with it, it's gone. So that leads to [big] hesitation by Big Pharma. So what

Big Pharma is doing right now is getting into personalized medicine through the diagnostic areas and focusing on acquisitions of drugs that have clear market value. Novartis has invested heavily in a personalized medicine plant that used to belong to Dendreon. Roche has a very strong internal program in personalized medicine. [In terms of made-to-order drugs] this is only for you, oncology can bear the economics. No one is going to pay $93,000 per year to treat their cholesterol. Economics make sense in oncology, whereas in most areas they wouldn't be based on chemotherapy, recovery time, lost time from work, etc.

Dvorak concludes by describing how made-to-order drug companies survive in niche markets. "(In terms of) marketing, Genitope used patient advocacy groups as our biggest marketing focus. We would go (to) those patient advocacy groups for those particular diseases, and make them aware of the clinical trials. We would generate some buzz and enthusiasm for people who were specifically looking for this type of treatment for themselves and their families. So they would go to their physicians, and the physicians would look into it. We would also market to physicians who were particularly focused in those treatment areas."

Thus, in terms of economics, Big Pharma and biotech are mainly invested in made-to-stock drugs, while foregoing made-to-order drugs for financial and regulatory reasons. Novartis makes Gleevec and Genentech makes Herceptin, for example, which are certainly very profitable for these companies. There exist a handful of made-to-order drug companies that make specifically tailored and personalized medicine for each individual patient, such as the former Genitope and Dendreon. Yet personalized medicine has been shown to be cost-effective in its companion diagnostics model of health care, and the cost utility of genetic tests is rising. Overall, personalized medicine will constitute a disruptive innovation that surely will have profound economic effects for society.

REFERENCES

Akhmetov, I., and R.V. Bubnov. 2015. Assessing value of innovative molecular diagnostic tests in the concept of predictive, preventive, and personalized medicine. *EPMA Journal* 6:1–12.

Caesar, M. Pharmaco-economic aspects of personalized medicine and companion diagnostics. Easton Associates.

McDougall, G. 2008. PriceWaterhouse Coopers Personalized medicine and its impact on the economy. *Population Health & Disease Management Colloquium*. Philadelphia, PA.

Davis et al. 2010. The microeconomics of personalized medicine. Washington, DC: McKinsey & Company. Available from http://www.mckinsey.com/industries/pharmaceuticals-and-medical-products/our-insights/the-microeconomics-of-personalized-medicine.

Paci, D., and D. Ibaretta. 2009. Economic and cost-effectiveness considerations for pharmaco-genetic tests: An integral part of translational research and innovation uptake in personalized medicine. *Current Pharmacogenomics and Personalized Medicine* 7(4):284–296.

PriceWaterhouse Coopers. 2008. The new science of personalized medicine. New York: PriceWaterhouse Coopers. Available from http://www.pwc.com/us/en/healthcare/publications/personalized-medicine.html.

Trusheim, M.R., and E.R. Berndt. 2015. The clinical benefits, ethics, and economics of stratified medicine and companion diagnostics. *Drug Discovery Today* 20:1439–1450.

Trusheim, M.R., Berndt, E.R., and F.L. Douglas. 2007. Stratified medicine: Strategic and economic implications of combining drugs and clinical biomarkers. *Nature Reviews Drug Discovery* 6:287–293.

Waldman, S., and A. Terzic. 2012. The value proposition of molecular medicine. *Clinical and Translational Science* 5(1):108–110.

Ward, M.M. 2012. Personalized therapeutics: A potential threat to health equity. *Journal of General Internal Medicine* 27:868–870.

Wong, W.B., Carlson, J.J., Thariani, R., and D.L. Veenstra. 2010. Cost-effectiveness of pharmacogenomics: A critical and systematic review. *Pharmacoeconomics* 28:1001–1013.

CHAPTER 12

Moral and ethical issues: Claims, consequences, and caveats

We are going to make sure that protecting patient privacy is built into our efforts from day one.

President Barack Obama

Personalized medicine is not just about scientific research and biomedical advances. It should be easier for stakeholders to enter the field and ensure that patients receive personalized medicine care in the most ethical manner. Patients and providers will have changing roles as personalized medicine becomes implemented. Patients are coming into physicians' offices with a greater knowledge of genetic risk and its implications, knowing about actionable lifestyle prescriptions for reducing the risk of disease. They will have greater access to medical records, thus allowing for the improvement of treatment decisions through patient education.

Genomic privacy forms a dilemma. There is a right to protect patient data, but scientific and medical stakeholders would benefit from having access to patient data for research and clinical purposes. Privacy is a personal and fundamental right guaranteed by the U.S. Constitution Privacy Act of 1974, and includes

- Inherent limits on the First Amendment as a constitutional right to privacy
- Fourth Amendment rights against search and seizure
- Due processes clauses of the Fifth and Fifteenth Amendments

Mark Gerstein of Yale Computational Biology describes this dilemma as the fundamental, inherited information that is very private versus the need for large-scale data sharing to enable medical research. If privacy is indeed a problem, then genetic exceptionalism concurs that the genome is potentially very revealing about one's identity and characteristics. Other risks may be revealed during a study for another trait. Alternatively, study participants might not even care about their genes, to make genomic privacy a conundrum. According to Gerstein, there are tricky privacy considerations in personal genomics. Some genomic information may be meaningless now, but not in 20 years or later down the line. Can true consent be possible when children are considered? Once put on the web, it cannot be taken back. Genomics supports open data, but it is not clear if personal genomics fits this paradigm. There is also the case of HeLa: what if your genetic data could rise to a product line, and what does this mean for ownership of data and consent? However, sharing helps speed research. Large-scale mining of this information is important for medical research, and sharing is important for reproducible research and useful for education.

Laura Lyman Rodriguez of the National Human Genome Research Institute (NHGRI) notes that identifiability is a shifting spectrum

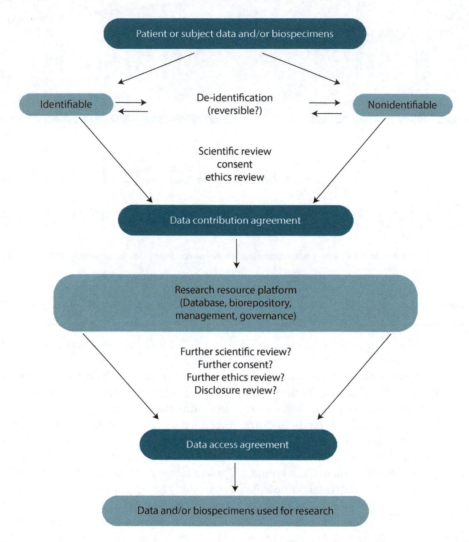

Figure 12.1 How patients will be deidentified as their data is used for research during the PMI. (Reprinted with permission from Lowrance, W.W., and F.S. Collins. 2007. *Science* 317:600–602.)

and has different definitions (Figure 12.1). There are a variety of means to render data identifiable, but still we must balance scientific potential with public trust and participant protection, and there is a natural tension between the value to protect and respect participants and promoting health advances through research. In a study conducted by the Institute of Medicine in 2009, 48% of participants preferred for each study seeking to use their data to contact them in advance and get their specific consent each time, with 24% preferring to have consent not needed if identity will never be revealed if the study is institutional review board supervised. Only 16.5% of participants would not want researchers to contact participants or use

their data under any circumstances (Institute of Medicine [U.S.] Committee on Health Research and Privacy of Health Information, 2009). In a study published in 2012 in *Public Health Genomics*, participants show deference to benefit over privacy concerns when asked to choose (Oliver et al., 2012).

Privacy will remain central to the Precision Medicine Initiative (PMI) (now currently known as the All of Us Research Program). According to Lyman Rodriguez, privacy and trust principles are the first steps in guiding all PMI activities, while articulating a core set of values and responsible strategies for protecting privacy and engendering public trust as the foundation for PMI. An inter-agency working group convened in March 2015 by the Office of Science and Technology Policy will draft the principles of privacy, particularly surrounding the cohort of 1 million volunteers. The PMI will be structured with participants as partners rather than subjects. The Foundation for the National Institutes of Health conducted a recent exploration of attitudes for PMI during 2015, sampling 2601 U.S. adults with an overall response rate of 57%. When asked which aspects of the study they would want to be involved in, the categories that received the highest ratings were to help decide what kinds of research are appropriate (62%), help decide what to do with the study results, (58%) and help choose what research questions to answer (Precision Medicine Initiative Working Group, 2015). According to a 2015 survey from the *Journal of the American Medicine Information Association*, 47% of respondents would be very likely to give consent to hospitals, with colleges and universities, federal agencies, and biotech and pharma trailing. Respondents would be less likely to give consent to insurance companies (Dye et al., 2016).

Healthcare providers will experience changing roles. Physicians will be seen as managers of care rather than repositories of all medical knowledge. They will rely more heavily on healthcare information technology for decision support, with improved care through use of aggregate patient data and highly networked team-based care. In addition, the pharmaceutical industry and insurance companies (payers) will see increased demands being made upon them.

Caveats will accompany and qualify these changes for individuals in terms of privacy of information and the possibility of genetic discrimination. Ethical and legal issues are associated with genomics, from its nascence in the Human Genome Project to the development of personalized healthcare, and throughout the entire life cycle of "base pairs to bedside" and "helix to health." The Genetic Information Nondisclosure Act of 2008 (GINA) prohibits group and individual health insurers from using genetic data for determining eligibility and premiums, insurers from requesting that the insured undergo genetic testing, employers from using genetic data to make genetic decisions, and employers from requesting genetic data about an employee or family.

The patenting of genes has also posed an ethical problem that the Supreme Court has faced. According to the Bayh–Dole Act, the function of a patent is to encourage investment in research and development by giving the patentee an initial period of exclusivity. When a patent application publishes the invention, it becomes public knowledge, and when the patent expires, it is available to the public and encourages others to design around it. A patent allows the owner to prohibit others from making, using, or selling the invention, resulting in royalties. Patent protection extends for 20 years from the date of filing. Natural phenomena, laws of nature, abstract ideas, and natural living organisms cannot be patented, but applications of the above can. There were a number of relevant cases prior to that of Myriad Genetics, including *LabCorp vs Metabolite* and *Mayo vs Prometheus* (2012), which held on to the patentability of a medical device. A 2013 lawsuit challenged BRCA1/2 patents, where the plaintiffs were the Association for Molecular Pathology and the American College of Medical Genetics, researchers and clinicians, genetic counselors, and patients, and the defendants were the U.S. Patent and Trademark Office and Myriad Genetics. Justice Thomas, writing for a unanimous Supreme Court, found that DNA isolated from its natural state is not sufficient to be eligible for a patent because it is still a product of nature. However, cDNA is patentable. In a subsequent ruling, the court held that primers are not patentable.

Ariosa Diagnostics vs Sequenome, in October 2013, was the first case applying the *Myriad* decision by a trial court. Sequenome is the exclusive license of a U.S. patent for a noninvasive method of detecting Down's syndrome in fetuses without the risk of inducing a miscarriage. Ariosa claimed that Sequenome's claim on the patient was invalid, and that cell-free fetal DNA is a natural phenomenon. Sequenome countered that the claimed methods were patentable because they are novel uses of a natural phenomenon. The district court concluded that the claimed processes are routine, conventional activity and not patent-eligible subject matter.

With these decisions in mind, David Resnick of Nixon Peabody Law Firm recommends including claims that are narrow, specific, and encompassing particular assaying and detection steps when filing a patent, and to tell a story (study) how you solved the problem. Consider carefully what you should patent and what should be protected by trade secrets.

ETHICS OF PERSONALIZED MEDICINE

Along with the opportunities provided by personalized medicine comes a host of obligations to serve patient care. The ethics of personalized medicine involves using new technology to achieve the fundamental moral purpose of medicine: good health outcomes for patients. Ethical issues in large-scale genetics and genomics research and individual whole genome analysis involve (1) vulnerable populations and the meaning of race, (2) biobanking, (3) returning research results, (4) enhanced oversight of genetic tests, and (5) direct-to-consumer genetic tests (Haga et al., 2013).

The principles of medical ethics are beneficence (doing good), nonmaleficence (avoiding harm), justice (making sure to provide the appropriate distribution of benefits), and autonomy (respecting the rights of individuals to make choices). Research integrity requires conducting research in accordance with standards that make it replicable and helpful, with the clear intention of answering questions and providing solutions, not a study that will waste people's time.

In terms of ethics, key objectives in pharmacogenomics must ascertain how the association between variant and drug response yields predictive value and improves the safety and efficacy of treatment. Pharmacogenomics must also assess outcomes from that type of information. Variants look at the breadth of information in the genome: there are 63 known pharmacogenomics variants and six novel variants of potential significance (Ormond, 2010). Thus, using genomes is an enormous asset to understanding genetic contributors to drug response.

The first huge ethical concern arises from the need for large databases to collect, store, and transmit information. There needs to be sufficient electronic capacity to study gene–drug interactions for both biological samples and detailed clinical and personal information available for data sharing. The National Institutes of Health (NIH) data-sharing policy states that

NIH reaffirms its support for the concept of data sharing. We believe that data sharing is essential for expedited translation of research results into knowledge, products, and procedures to improve human health. The NIH endorses the sharing of final research data to serve these and other important scientific goals. The NIH expects and supports the timely release and sharing of final research data from NIH-supported studies for use by other researchers (Final NIH Statement on Sharing Research Data, February 26, 2003).

One issue with the use of data relates to informed consent, the process for obtaining permission before conducting a healthcare intervention on a person. Many times, consent involves the patient signing a form that details the protections and rights of the patient during the treatment or clinical trial. Data banks make use of informed consent, since informed consent can be obtained for broad use, and this has been accepted by society as well as by most bioethicists. However, informed consent assumes the patient's right to decide to participate based on full knowledge of the study. According to bioethicist Wylie Burke of the University of Washington, this is not feasible for data repositories because data may be used for purposes beyond the patient's knowledge. One of the primary challenges will be to enhance and expand protections, given the limited role of informed consent through meaningful preauthorizations, communications, consultations, and delegation of oversight to an appropriate body to allow for reconsent.

The issue of data collection for genomic studies was highlighted in the examples of the sharing of data from two tribal communities, the Havasupai and Nuu-Chah-Nulth.

The Havasupai tribe consented to have their genomic data used for diabetes research that would potentially benefit the tribe. However, the researchers in question used blood obtained from tribe members for research on diseases other than diabetes, such as mental and cardiovascular disorders, and were not up front with the tribe about their research objectives, as the tribe claimed. The Havasupai countered that the researchers did not care about what would benefit their tribe, but only about their research. The issue here is not whether the research was done correctly, but that the researchers did something they should not have done.

Another case concerning the ethics of data sharing involves the Nuu-Chah-Nulth tribal community in British Columbia, Canada. According to Ron Whitener, who has written a white paper on Native American tribes and genetic issues they have faced:

A case study related to data sharing for secondary data analysis involves the use of tissue samples taken from members of the Nuu-Chah-Nulth First Nation of British Columbia, Canada. In the early 1980s, tribal members gave samples to a researcher from the University of British Columbia to study rheumatoid arthritis, an ailment common to their aboriginal community. The researcher was a well-known geneticist whose research included the study of associations between race and genetic disease and population genetics. Research on rheumatoid arthritis was conducted and findings were published. When the researcher moved from the University of British Columbia to the University of Utah, and then to Oxford University, he took the samples with him, used them himself, and loaned them to other investigators for types of research not covered by the informed consent agreements signed by the Nuu-Chah-Nulth informants. When the Nuu-Chah-Nulth discovered that research was continuing with their members' tissue samples, they demanded that the tissue samples be returned to the community. The result had a chilling effect on research in that community, with the chairman of the Nuu-Chah-Nulth Research Ethics Committee stating "hopefully the day will come when we can put it all behind us, but we've unfortunately learned a lot about the dark side of scientific research…. Our eyes are wide open now." The researcher later expressed understanding at indigenous communities' concerns and stated his wish to return tissue samples he had collected. While the Nuu-Chah-Nulth were angry at what the researcher had done, the fact was that the practice of keeping samples obtained through consents given for a specific study purpose and reusing and sharing with other researchers was considered standard practice at the time in Canada. Canada later revised its policy and required consent by the respondents for any future use of data containing individual identification. The new rules had an exemption for this requirement if the researcher could show that the research cannot be achieved without the personal information, it is impracticable to obtain consent, adequate safeguards are in place to protect the information, and the Research Ethics Board has weighed the public interest in research versus the public interest in protecting individual's privacy (Whitener, 2010, unpublished paper).

Other studies on data sharing illuminate the need for researchers to protect participants in terms of asking permission before submitting data for sharing and ensuring that participants are made aware of what researchers are doing with the data. Participants prefer research updates because they donated something of themselves toward a project and expect to learn the results of the project.

In short, matters of trust and expectations exist on the part of participants. The cases of the Havasupai and Nuu-Chah-Nulth raise questions about how to plan data sharing. Participants want to make sure research is sound, but they also want to know how their samples are being used and assume they will receive information (research results) beneficial to them.

In addition to addressing the subject of informed consent, there is the need to acknowledge the essential role of the public in funding and participating in data-sharing projects. According to Burke, "we must develop strategies to keep participants informed and stewardship of shared data resources must be made public."

Another ethical dilemma associated with personalized medicine and genomics concerns whether tests improve outcomes. In prescribing drugs, pharmacogenomics variants are screened for. Technology panels are currently in use, but they are giving way to targeted gene sequencing, whole genome sequencing, and whole exome sequencing. According to Burke, the community should verify that pharmacogenetics testing is more clinically effective than alternatives.

Variants can be grouped according to penetrance, from highly predictive variants with high penetrance (patients who have the variant will get the disease) to weakly predictive gene variants with low penetrance (some patients will manifest the disease and some will not). How can information about penetrance of variants be used to decide if whole genome sequencing is the correct way to proceed? What information can we pull out of it and why? Where is the cutoff point, and how do we decide what is useful for people and who decides?

There is a cascade effect of medical technology. Incidental, ambiguous, false positive results lead to further tests and adverse effects. The questions remain: When is whole exome sequencing and targeted sequence testing the right test? Under what circumstances does genomic medicine help?

Delivering the benefits of personalized medicine requires high-quality practice guidelines, relevant evidence for the clinical setting, and population and outcome data indicating access for all. The distinction between disease and prevention must be made, since there are different criteria for both. If one has a disease with a defined course, then coming up with treatment options would have a very different risk profile versus finding out about an illness from a preventive standpoint. According to Burke:

> The criteria for decisions about healthcare should be recommendations that are morally justified and can be explained to the public, with alternative views respected, driven not by opportunism, personal interest, or technological imperative, but by a stream of evidence of how we can achieve the [appropriate goals for personalized medicine]. It is not an easy process.
>
> Clinicians have an obligation to understand technology they are using, and use in their patients' interests. There are only narrow ways genomics can do this. [There are issues of] resource allocation. Is expensive genomics the right way to spend Medicare dollars? We are not ready for whole genomes for everybody. It's not cost effective—not to detract from research on genomics. The biggest gain from genomics is not developing genetic susceptibility

profiles, but in understanding disease at a molecular level and better surveillance to catch people early. Genomics at the public health level far more explicates disease than as a test for individuals.

What results should be turned in and how should it be used? [For example] how will clinicians help mental health? This is a big question mark. Genetic disposition can predict depression and schizophrenia, and pharmacogenomics can prescribe drugs more effectively, but [to] what extent will gene variants for mental health confer stigma? Regulations and laws form a floor, a bottom line, the minimum we need to have. Bioethics [concerns] how we can be more responsible. The question for regulation: Is our floor solid? We do have GINA.

Hank Greely, JD, a bioethicist and director at the Center for Law and the Biosciences at Stanford Law School, cites the accuracy of whole genome sequencing as a progenitor of potential problems associated with personalized medicine. Whole genome sequencing does not count the number of repeats of a DNA sequence, which is crucial for Huntington's disease. Whole genome sequencing is also not very good at detecting small deletions and insertions, nor translocations (however, this is rapidly changing). It is also not good at figuring out which chromosome variations are on; if a recessive disorder has both alleles on one chromosome, the person is not affected. If both alleles are on different chromosomes, the individual is affected. Since the accuracy of whole genome sequencing is not perfect, this will make a difference in terms of how accurately they are reflected in different regions of chromatin; for example, regions with high cytosine–guanine ratios and lots of heterochromatin. In these cases, accuracy is going to be crucial.

Positive predictive value demonstrates how complicated the concept of accuracy is. Let us say one gets a positive test result; how common the underlying trait is in reality dictates whether a positive test is true or false. For example, if there is a sex determination test that determines whether a child is a boy or girl, and there is one false positive (51 boy results; 50 boys and 1 girl in reality), the positive predictive value is good. Positive predictive value becomes more complex as we move into the whole genome sequencing world, with variation inside genes. Every time we sequence someone, we see tens of thousands of sequence variants. Then, we must ask how common the variant is in the population. One in a thousand? One in ten thousand? The person may be the only one in the world with the variation. How good does specificity have to be on a test to give it a good positive predictive value? To take another example, let us say a test makes a mistake every one in a million times, but only one person in a billion has the variation. Thus, the test has a poor positive predictive value. In short, how accurate a sequence is matters when looking at uncommon variations, and we will need to know for each sequencing machine how good is it after 3 months, 6 months, or even 12 months. According to Greely, we need to establish rules for maintaining the accuracy of testing equipment.

Greely also asks how we are going to be sure that the sequence we get is the right one. Will we go back and double-check our results? Let us say we have a whole genome sequence that indicates a potentially pathogenic problem in this particular gene. Do we revert to the Sanger sequencing method, or do next-generation sequencing again on another machine? In any case, we must distinguish true positives from false positives. Which is the most reliable and cost-effective way of checking the sequence? How do we deal with false negatives? Rechecking negatives means rechecking whole genomes. Do we recheck on two different machines with two different strengths and weaknesses? According to Greely, we have to design protocols to determine how accurate the information should be. Clinical Laboratory Improvement Amendments (CLIA) will play an important role as a clinical laboratory in maximizing the accuracy of actual sequences. As Greely says, "If the sequence is no good, the results are no good."

Greely also states that the accuracy of interpretation is very hard. If we assume we have a completely accurate sequence that is neatly aligned on all chromosomes, and we can see all of the translocations, and we can be confident in what that sequence is, what does the sequence mean to the patient? According to Greely, that depends on the interpretation. There are a number of factors involved in interpretation. First, who is going to interpret it? One possible answer is to attract different competing private firms to the sequence interpretation business. These firms will send back the interpretation to the consumer or the physician indicating what the sequence means in terms of risk this patient bears. Imagine 10 different companies conducting the interpretation and not agreeing on one interpretation. We could also have public or nonprofit companies providing interpretations. Even in that scenario, there will again likely be differences among interpretations.

We could alternatively designate an entity to be a "monopoly interpreter." However, according to Greely, we do not know how that will work out. We need somebody to fund an open process constantly in a consensus conference where people are regularly looking at curating the new papers that come out with respect to each gene or each disease, regularly reading them, making tentative conclusions, putting those conclusions out on the Web, and soliciting public comment on them. Greely's vision is a curated wiki-like process, making a set of recommendations about those interpretations, but they are recommendations that are transparent. An interpretation could answer why the interpreters think this variant has pathologic significance but another does not confer disease. And there again can be differences among variant interpretations. "This would be a good way to move forward with how we interpret whole genomes." Greely is optimistic that we will reach a situation where one entity or several entities will have software that will plug the genome in at one end and out from the other end comes a set of findings with red flags, yellow flags, and green flags. These set of findings would then be sent off to clinicians or patients or consumers for their interpretation or use. Computers would do the main work, with recommendations sent off to the clinician; the computer software will be a medical device regulated under the Food, Drug, and Cosmetic Act and regulated by the Food and Drug Administration (FDA). How the FDA will regulate this software remains a question; we will have to invent a new regulatory scheme so that these interpretive packages actually are safe and effective.

We must also consider the issue of returning the information from whole genome sequencing. How do we return the information, which information do we return, and to whom?

First, to whom do we return it? Greely comments on direct-to-consumer genetic testing: "We cannot have direct-to-consumer whole genome sequencing; the potential for disaster is too high. People will need help in dealing with the complexities of the genome that affect something as emotional as one's own health and the health of family members. We are going to need expert guidance in order to make sense of it." Greely worries that if we just give whole sequence genome data or interpretations back to individuals, "they will screw up because they are human. People will make bad decisions." For example, if compared to the population at large, for a two to three times increased probability of developing a disease, you may still be at under 50% risk of developing it.

Another example is one where a woman gets back a test with the good news that she does not have the BRCA1 or BRCA2 mutation. Thus, she is not one of those women who are at a 55%–85% risk of getting breast cancer. She may decide not to have mammograms, which would be a bad decision. Not having the mutations means she has gone from the 55%–85% risk to the nationwide risk of 12%–11.97%. "People (don't have a good understanding of) percentage and probabilities. We need help." Technically, whole genome sequence information should be the result of restricted devices, and the FDA should only allow them to be used with a doctor's prescription, and the results should be only returned to the doctor, with the doctor taking responsibility to help the patient make sense of the results.

Another question remains: What should be returned? The biggest fight today in research ethics is about whether incidental findings from research should be returned to research subjects. Greely is of the opinion that for anything that is important clinically, there has to be an obligation to share it with the patient. Let us say a patient is being tested for a gastrointestinal disorder, and the whole genome sequencing reports no gastrointestinal disorder but does show the presence of a sudden cardiac death gene. Greely feels that we have an obligation to tell the patient about the sudden cardiac death syndrome gene. "It's a fiduciary obligation to that patient. If one has an x-ray for concerns about liver but the doctor observes a big tumor in the lung, the doctor automatically tells the patient about the tumor in the lung." However, "if the patient doesn't want to know, and really doesn't know, and the clinician sees something really serious, highly penetrant for which there is a good intervention, I am going to doubt the patient doesn't want to hear (about it)."

The vast majority of existing biobanks do not have resources to establish fund processes to return research results.

How are we going to tell the information to the patients? How are we going to train physicians about 6000 genetic diseases and 40,000 genetic associations? How physicians get trained is of paramount importance. How are we going to convey this information to patients? Who is going to be able to explain what all this information means, and who is going to listen? According to Greely, "We need to be able to invent new ways to get information back to patients. These ways should still incorporate (a) face-to-face meeting with a skilled person." An in-office meeting with a genetic counselor is still more meaningful, but whole genome sequencing offers too much information to convey all of it face-to-face conveniently. Greely supports investing in Internet and video as a means of doctor–patient communication, which would allow doctors to "look at whole genome sequencing results in advance, engage the patient, and tell them in advance what they should be worried about, and alert them to three or four things the genetic counselor wants to talk about, but give patients the opportunity to talk about other things too. We need new ways of conveying information."

How often do we return information? A patient's genome might not change, but understanding of that genome may change daily or weekly as new studies confirm (or fail to confirm) or discover different associations between disease and variations. Greely states, "If we never reinterpret the genome, we are not doing well by the patient. We need some sort of standard of care, to run the genome through software again, in six weeks? Six months? Six years? We don't know what [the] number will be." This depends on how important genomic inform will be and how rapidly it changes. "Genomic diagnosis can't just be a one-off thing because information changes. It leads to other questions. What if the patient relocates, what is the physician's obligation to try to find him to give him this new information."

Now we ask who is going to hold the information: The patient or the primary doctor? The sequencing center or the interpreters center? Are they all going to hold it? Each additional party who holds the information poses an additional source of problems for leaks, and yet also provides an additional way of finding and connecting that information with the patient. Greely envisions a centralized holding tank, such that if the patient relocates, or if a thumb drive with the genome sequence is lost, the genome can be retrieved and reinterpreted and reapplied to the patient's case by his or her new doctors. Who is going to hold the information also concerns confidentiality. Once the genome is available on a thumb drive or in the doctor's office, confidentiality is at risk. If the data is somehow accessible, police can get to it, employers can get it. According to Greely, it is fully protected as all other medically information, which means "it's not well protected."

Children offer a special situation in the personalized era. The standard has been not to test a child for a genetic illness if preemptive action cannot be performed on the child. For example, a child should not undergo a test for Alzheimer's. However, Greely asks when we have whole genome sequencing at birth, what do we do with that sequence? What should parents do about it? Not open it until the child turns 18?

Overall, Greely is an optimist. "Genomically based medicine will be a good thing, leading us to understand disease better, coming up with better preventions and treatments for disease. (It will) help people adapt their lifestyle and environment in ways that work better with particular genomes. It's not going to make death go away, but it should make good incremental improvements. But things could also go bad, we need to worry about legal and ethical issues and practical steps with accurate effective and safe tests and ways of conveying that to people. Or we risk being in worse shape (than we are now)."

Gloria Peterson of Mayo Clinic Health Sciences offers a researcher's perspective on incidental findings and biobanks. She describes the biobanker's perspective on banking patient samples and where biobankers stand. Her conclusion is that standardization is needed in the field of complexity. Biobanks do not have the responsibility to return incidental findings and are not legally obligated to retain identifiers or contact information of subjects who contribute samples. She contributed to a study finding that there are new responsibilities for biobankers: findings that are analytically valid, reveal an established and substantial risk of a serious health condition and are clinically actionable should generally be offered to consenting contributors. Peterson notes that her study has met with rejoinders.

Thus, a number of bioethicists have shared their insights on the moral and ethical consequences of the Human Genome Project and personalized medicine. As personalized medicine moves forward, issues such as data sharing, the ethical use of genomic data, confidentiality, whole genome sequencing accuracy, incidental findings, and patient privacy will remain paramount. In conclusion, the outcomes of these ethical caveats on the claims of genomics and genomic medicine remain to be seen.

REFERENCES

Dye, T. et al. 2016. Sociocultural variation in attitudes toward use of genetic information and participation in genetic research by race in the United States: Implications for precision medicine. *Journal of the American Medicine Information Association* 23:782–786.

Haga, S.B. et al. 2013. Public knowledge of and attitudes toward genetics and genetic testing. *Genetic Testing and Molecular Biomarkers* 17:327–335.

Institute of Medicine (U.S.) Committee on Health Research and the Privacy of Health Information. 2008. The HIPAA privacy rule. In *Beyond the HIPAA Privacy Rule: Enhancing Privacy, Improving Health through Research*, Nass, S.J., Levit, L.A., and Gostin, L.O., eds. Washington, DC: National Academies Press. Appendix B: Commissioned Survey Methodology. pp. 1–26. Available from http://www.ncbi.nlm.nih.gov/books/NBK9583/.

Lowrance, W.W., and F.S. Collins. 2007. Identifiability in genomic research. *Science* 317:600–602.

Oliver, J.M., Slashinski, M.J., Wang, T., Kelly, P.A., Hilsenbeck, S.G., and A.L. McGuire. 2012. Balancing the risks and benefits of genomic data sharing: Genome research participants' perspectives. *Public Health Genomics* 15:106–114.

Ormond, K.E. 2010. Challenges in the clinical application of whole genome sequencing. *Lancet* 375:1749–1751.

Precision Medicine Initiative (PMI) Working Group. 2015. *The Precision Medicine Initiative Cohort Program: Building a research foundation for 21st century medicine*. Bethesda, MD: National Institutes of Health.

Whitener, R. 2010. Research in Native American communities in the genetics age: Can the federal data sharing statute of general applicability and tribal control of research be reconciled? Unpublished paper.

Conclusion

Every day matters for the patient.

Paul Hudson, MD
President, AstraZeneca US

Geisinger Health Systems serves as a final example of personalized medicine in action. David Ledbetter, PhD, chief scientific officer, describes how their biobank began in 2006 in a 12-year longitudinal study to sequence 100,000 Geisinger patients. Signed with Regeneron, this study elicited a broad return of results with high evidence-based gene pairs. Every BRCA carrier in the population has been found. Additionally, their results revealed at which age statin treatment should begin for individuals who had familial hypercholesterolemia to prevent heart disease. Low in-and-out migration families and third generation families who have implemented electronic medical records have undergone a 12-year longitudinal study resulting in data and biobanks. With a 2006 start date with a unique trust rate and high consent rate, this constitute the perfect place to conduct longitudinal genomic research. This type of collaborative venture is seen as an exemplar of the clinical implementation of personalized medicine that can led to actionable results.

According to Dietrich Stephan, PhD, founder, CEO, and president of Silicon Valley Biosystems, for personalized medicine to work:

[we need a] scalable infrastructure. Precision medicine's mission is to enable the clinical community to bring the genome to the point of care to improve health globally, inexpensive and scalable, and facilitate accurate diagnosis for precision medicine. [We need the] sub-1000-dollar genome for personalized medicine to be pervasive. When we sequence at birth, we see value across lifetime.

The value of the genome resides in electronic medical records. There is room for software-based genomic diagnosis return [in tablets returned to physicians].

However, enormous challenges for the implementation of personalized medicine persist. Matt Posard of Illumina, a company that is a major player in the next-generation sequencing field, explains, "Medical community engagement (remains an issue). In a survey, 98% of physicians gave a thumbs-up for genetic diagnosis, but only 10% actually administered a genetic test, even though 63% believe genetic testing gives physicians the ability to give a better diagnosis." The cost of test reimbursement is also unresolved. "The potential (to reimburse) is not big enough to reimburse for genetic interpretation services; we need cost savings. (It may be) easy for physicians, for patients, or for genetic counselors, but we need clarity from the (pharmaceutical and biotechnology) industry and FDA (Food and Drug Administration) in terms of how."

According to Dan Roden, there exist a number of finite and concrete challenges to implementation of personalized medicine.

The challenges to implementation are multiple. One challenge is to develop the basic outcomes dataset that will allow you to say that this particular variant exerts an important effect on a drug response, and that would be one observation you would have to have some confidence in. You need to know the source of variability, what action you can take to overcome that, and what happens when you take that action. Does treatment work? Building the evidence base is one great challenge. Part of that is not only understanding the evidence, but how big of an effect the genetic marker has. Does it increase the likelihood of drug working by 10%, or does it increase the likelihood of a drug having a fatal reaction by a million percent?

Once you understand that part [and we are far from that for most drugs], trying to understand the way in which that information will be deployed clinically is the second challenge. One way of doing that is to educate physicians and healthcare providers that particular genetic variants are important in outcomes for particular drugs, that the recommended course of action is to mitigate that variability [increasing the dose or changing the drug]. The alternative approach is that physicians don't need to know that information. It is a robust healthcare system that says this drug has now been prescribed, we have on file this patient's genetic information and the physician will act on that as soon as the drug is entered. The system then responds by saying that that patient had genetic testing and this dosage you prescribed is too large for that individual, for example. [Vanderbilt] [where Roden is professor of medicine and pharmacology] has had the option of listing that kind of advice to develop mechanisms to assist the healthcare system in responding to those variants.

The third challenge is educating physicians on how to interpret the data [when to take the warning seriously or when not to]. A fourth challenge is tracking outcomes, or how to get data into the patients' records in a way that will allow the data to be used at some future date. Another challenge is knowing how sure you are the genetic data you are being given is correct. Every time you do a test, there is a finite possibility of an error. For some lab tests, it is pretty obvious to a clinician that the result is an error. But if you are getting genetic information, there is no way to look at the genetic information and say, "I don't think that is the case, we should repeat the test." The genetic information is relatively privileged and you have to make sure you get the data right the first time. Those are among the major challenges.

Personalized medicine becomes very relevant upon the realization that every human disease is genetic to some degree, and genetic testing becomes relevant to every single human disease, particularly monogenic disease (of which there are 4000) or hereditary cancer mutations that are inherited from parents. These disorders and mutations are highly penetrant, and we need to understand how to diagnose them. Why is genome testing unique for diagnostics? Mutations have unique characteristics from carrier testing to fetal cell testing. One test can diagnose every Mendelian disorder, which obviates the need of going through family history to determine which disease to test.

While monogenic disease constitutes 5%–10% of diseases globally, chronic diseases such as diabetes, cancer, and hypertension comprise 80% of diseases globally. According to Stephan, we need to understand precision prevention to stratify populations in terms of modifying environmental triggers and how they impact healthcare. We need to flesh out gene–environment interactions.

Other obstacles present themselves. Nobody bought the Roche P450 chip, which individualized metabolic differences in patients, underscoring the inability of some personalized medicine advances to meet patient markets. For all the cancer therapy that results from cancer pharmacogenomics, including the layering of targeted therapy or made-to-order drugs with conventional treatment such as radiation and chemotherapy, there is no precedent or any evidence-based medicine to support it, although opportunities remain. Researchers are also aware that the microbiome interacts with the germline genome.

Personalized medicine is contributing to the research front in the following way: through the translational research lab, scientists ask doctors what they need, go into the lab, and create what the doctors need. However, as Stephan notes, "nothing that scientists did went back to doctors." Stephan offers the idea to set up a biocluster translating genomic discoveries into treatments for physicians and patients, since this is effectively blocked.

Edward Abrahams of the Personalized Medicine Coalition comments that we are "not invested in this paradigm as we should be. The NIH (National Institutes of Health) cuts served personalized medicine very badly, (funding that would have helped) in harnessing the power of understanding the human genome." In the Centers for Medicare & Medicaid Services (CMS), there is a movement to contain costs. Abraham continues, "The problem is (that) the (CMS) doesn't understand the promise of personalized medicine. They think personalized medicine will lead to increased costs and will cut reimbursements for molecular diagnostic tests. If this happens, this will limit patient access and the next generation of diagnostics that would be developed." Essentially, we are missing the delivery of the genome to the diagnostic point of care.

At Navigenics, a personal diagnostic company that went bankrupt in 2009, researchers took a representation of the genome and provided actionable information for chronic disease risks and pharmacogenomics. They added monogenic disorders, cancer pharmacogenetics, and host–pathogen interactions. However, according to Stephan, "They were too early."

There also remains a problem with whole genome sequencers. At the sequencer stage of the technology adoption curve, the technology is still not broad-based care in community settings. Through sequencers, researchers and clinicians grapple with assays, extract patients' variants to find polymorphisms and mutations, annotate them, classify those which are relevant to the clinical trait of interest, and then report results and act on them. In terms of variant detection concordance, it is very difficult to find variants for clinical disease. Which one do we pick out of the many? We need a benchmarking engine that is similar to an Internet search engine.

In the case of variants of unknown significance and pathogenicity prediction, technology appears to be working well in terms of clinical grade sensitivity and specificity. However, the question remains of whether the variant causes disease. How do we put the final piece of the puzzle in place the next time the physician sees the patient and has the insight that the variant has clinical significance? This is the piece of the puzzle that, according to Stephan, fuels academic research. However, the community and everybody else remain outside of this research circle.

The TMN (tumor, metastasis or whether the breast cancer has metasized, node or lymph node) system for staging breast cancer is accurate at 60%; Genomic Health's Oncotype DX assay is changing the way we think about diagnostics; however, surgeons are hanging onto the TMN system when they should be turning to Oncotype DX to eliminate surgery and also know better how patients are doing.

Whole genome sequencing must become cheap and pervasive and improve patient outcomes, and its results must be actionable across all areas. This must be driven globally through a partnership with major medical centers. Physicians must get educated. Stephan notes that the "key is specialty societies who come up with recommendations that are listened to. We have experts talking to experts. University of California and Stanford research is not really clinical; however, clinical people have to make it happen."

Paul Hudson, president of AstraZeneca US, encapsulates the history of precision medicine. Six years ago, an important breakthrough was made with the discovery of a targeted mutation for lung cancer. According to Hudson, the precision approach consists of 10–12 months of the patient doing well on average, targeting the mutated receptor, treating the disease, treating the mutation, and thus treating the individual patient.

However, Hudson acknowledges that we need to understand why resistance emerges to therapy. Now patients are taking health into their own hands. After FDA approval of one of AstraZeneca's targeted therapies in 32 months, one of the fastest approvals, at 9:30 a.m. on a Friday, by the following Thursday, a handful of patients were initiating treatment.

Breakthrough collaborations for companion diagnostics with drugs, such as those between Eli Lilly and Abbott in the area of asthma, are occurring, supporting smart clinical trial design. In the twenty-first-century Cures Initiative, the FDA is using real-world data and being more broadly accepting of modified clinical trials. Barack Obama's Precision Medicine Initiative is meaningful policy; it will remove barriers and ensure revolutionary progress with innovation and redefinition of the value of care. Hudson adds that it will reduce the cost burden on hospitals, with an evolving pricing structure for patient affordability.

AstraZeneca is entering into an aggregate with payers in the area of risk sharing based on patient outcome, which is challenging in the current political environment. Many breakthroughs of tomorrow will be enabled by personalized medicine, including circulating tumor DNA, which will allow for repeated opportunities to test non-invasively. BRCA testing is up more than 60%. The business model of pharma is going to be challenged. The single biggest investment of the healthcare dollar is to show that money is being invested in the right way, and the pharma model is going to have to catch up. The industry will have to reinvent itself and show it has value working for patients.

In short, the implementation of personalized medicine, particularly in clinical community settings, faces many barriers, most of which concern data. Electronic medical records and information technology in general will also assist in overcoming obstacles, especially surrounding data. Partnerships between community physicians and community hospitals and major medical centers can overcome some of the barriers. Pharmacogenomics, with its discovery of drug–gene pairs through basic research, promises to revolutionize clinical care once translated from discovery. Medicine has come a long way from the days of Hippocrates, and the journey to better patient care is always imminent. In conclusion, by utilizing the information commons and knowledge network, companion diagnostics, and whole genome sequencing (among other technologies), personalized medicine and its branches are poised to advance healthcare and biomedicine tremendously. We should definitely be investing more in developing this paradigm.

Index